Up, Up, and Away

(Or; 'Naked into Truth')

Liz Mitten Ryan

Copyright © 2021 by Liz Mitten Ryan

All rights reserved. This book or any portion thereof may not be reproduced or transmitted in any form or manner, electronic or mechanical, including photocopying, recording, or by any information storage or retrieval system, without the express written permission of the copyright owner except for the use of brief quotations in a book review or other noncommercial uses permitted by copyright law.

Printed in the United States of America

Library of Congress Control Number: 2021901635
ISBN: Softcover   978-1-64908-818-5
        eBook   978-1-64908-817-8

Republished by: PageTurner Press and Media LLC
Publication Date: 01/28/2021

To order copies of this book, contact:
PageTurner Press and Media
Phone: 1-888-447-9651
order@pageturner.us
www.pageturner.us

Up, Up, and Away
(Or; 'Naked into Truth')
A light and entertaining romp through life, from confusion to absolute knowing.

Chapter One

EARLY YEARS

I remember the first shock, the feeling that everything was not as it should be; there had to be some kind of error; there it was plainly staring out at me, a big, black U, unsatisfactory in the behaviour category of my grade two report card along with some mediocre marks in academics.

"If Elizabeth would pay more attention in class, her marks would certainly improve."

"I got an 'A' in art," I proudly told my mother, trying to soften the blow. My mother was not impressed. She had always been one to rally for the institution. When in grade 6, the principal wanted her permission to give me the strap; she was all for corporal punishment.

Throughout the early fifties, I was a trendsetter, breaking rules that were not yet in place. Setting new parameters, always with a smile on my face that I was told to take off. That alone caused me self-introspection. How could one change their looks? It sounded like something dishonest.

Another honor I was gifted was in high school, after being the first girl in elementary school to get the strap, I became the first in high school as well. I was even offered my own special classroom in a coat cupboard just outside the principal's office. It had no windows, so I had to pass notes to commiserators between classes through the crack in the door. In spite of my mediocre marks through school, I was accelerated grades three through five as a result of a new IQ test that told them I was able to put the appropriate pegs in the appropriate holes very quickly. Accelerated meant doing three years in two and being ejected at the other end now, one and a half years younger than the majority of my classmates, which fostered a whole new set of injustices as far as I was concerned.

My parents kept to an unreasonably strict set of rules to keep me safe in the company of my more advanced cohorts, and I kept to a regimen of militant and devious behaviour in the name of fairness.

I can still see my poor fathers' blood pressure rising, in a red line up his confounded face as I declared my freedom, ran, and locked myself in the bathroom and refused to cooperate.

"Lizzie, come out of that bathroom!"

"No!" Lizzie, come out now, or I will take the door off the wall." Being an engineer, I knew he could do it, so I maintained my bravery until I heard the screwdriver at work, then shinnied out the window, down the roof, leaping the 5' fence and disappearing in the dark with my father in hot pursuit.

The other subject I was good at was gymnastics, and I remember feeling empathy mixed with remorse as I watched from a neighbors' hedge; my father circle the neighborhood in the dark, scanning with his car lights.

I spent a lot of time in my room with my stereo and my record collection, dancing and singing at the top of my lungs every word of my beloved musicals. I remember after one particularly bad report card, my dad cutting the cord off my stereo forcing me to study. Still, I'm not sure it worked, and I did learn how to re-tape the wires.

My other dream world was where my horse lived in the basement. I padded a big old bookcase with cushions and blankets, made a belt halter, and rode through the wide-open spaces (in my mind) stopping for campfires and meals. I had my bedroll and pillows and slept under the stars. I rode that bookcase, gradually training it to trot as it got looser in the joints, front and back, front and back, until one day it lurched forward, tossing me on the ground. Coming to the end of its life as a rideable mount, it was put out to pasture. On nice days I had my cowboy friends, and we ran the neighborhood with our six guns (complete with greenie stick em caps), chaps, and cowboy hats. A neighbor had the perfect tree branch just four feet off the ground that curved upward to become a neck. Out came the pillows and blankets, and a new horse was born. One day a real pony arrived on our block, a 'magnetic' prop for a photographer. It was decorated with a western saddle and breastplate. I think the owner was shocked when a whole posse of cowboys descended and surrounded them, but I'm sure we made his day.

 I also took pleasure in toying with the psyche of my brother Steve. He was four and a half years younger and owned a complementary set of genes. We were different, and of course, my version of life was correct... Everything in his room was in perfect order, his plant experiments on the window sill, avocados and grapefruit seeds, the hippopotamus, piggy bank with its open mouth, and its silver dollar collection, always perfectly arranged. I would sneak in and just slightly re-arrange things, even the silver dollars, so that different ones were on top. It made him crazy, and I found that very amusing, so he devised padlocks, and I found ways to jimmy the padlocks, again, just slightly re-arranging his collection and locking up behind me. It was very entertaining, at least for me. It was an innocent act, born of my twisted sense of humor that I think destroyed our relationship for life.

 It was his birthday, and I painted a portrait of him. Remember MAD magazine and Alfred E. Newman? He bore a shocking resemblance through my eyes, so I leaned in that direction. The scrawny 'long drink of water' he had for a body was in its birthday suit with nothing but a fig leaf (a small one) over its privates! Well, that did it. I waited until all his friends were present in a circle around us and unrolled my birthday gift to him, a ribbon tied scroll. Really,

I thought it was funny; he not so much, and I have actually heard reference to that moment decades later.

Who would guess?

I think possibly my confinement was getting to me as I was still rarely allowed to go to parties or sleepovers that my friends had. I did sneak out to a few, but there were consequences. In our house, we were mostly directed to the basement, which circumnavigated the house and had cement floors, albeit a 'wreck' room in one corner that my dad made for us. I mostly just let out steam there, wheeling around on roller skates, playing scream and chase games like 'monster.' Basically, a loud version of tag on wheels. My parents had block parties often where the adults were upstairs and six or so kids below. I really don't know how they were able to hear themselves think. The other quieter game was hide and seek, in which we took over every nook and cranny, every cupboard in the house, completely rearranging their contents. I think my mother had given up on any semblance of order as we would climb to the back and create piles in front to cover our presence.

We also went on picnics and ski trips to guest ranches and spas. (Think Schitt's Creek with younger kids and friends). They really were good parents. I was ever ready to inspire a laugh, eat, too much sugar and be entertaining (I thought) There were periods of solitary confinement and forced study.

"It just wasn't fair"...

Nothing seemed fair in those years, but somehow I managed to get through high school with my virginity and a C+ average that allowed me into university, which was much more accessible in those years, particularly since my curriculum was husband-hunting.

I should probably mention that I had always had conversations with God from a very early age. I would sit on the dark stairs behind the grill at the bottom of the door to the upstairs, where I had been banished for one reason or another and ask for guidance. I knew I was a good person with a pure heart, but somehow that was often lost in translation or clouded by my diligent fidelity to my individual truth. I knew I was a unique gift from God and could only be guided from that higher perspective. That guidance began to show itself early in the interesting and sometimes amazing circumstances that seemed to direct the thread of my life.

I adored my extended family and was proud of our heritage. We were good people, Puritans who had come over on the Mayflower generations earlier. In fact, our direct descendent was Peregrine White, the first baby born after the Mayflower landed at Plymouth Rock. I know this as my great Aunt Middo had remained in the States and was a member of the Mayflower Society and the DAR (Daughters of the American Republic).

Our claim to fame was our link to several times great, Uncle Benjamin Franklin who was known for myriad things from flying kites and almost being electrocuted to some of his many quotes like "In wine, there is wisdom, in beer, there is freedom, in water, there is bacteria."

"Beer is proof that God loves us and wants us to be happy."

And some famous ones like

"A penny saved is a penny earned." "Early to bed and early to rise makes a man healthy, wealthy, and wise." I resonated with this one:

"If you would not be forgotten as soon as you are dead and rotten, either write things worth reading or do things worth writing."

Among his inventions were the lightning rod, bifocals, and the Franklin Stove, and his likeness is still on the largest denomination of American money, bestowed as an honour, the hundred dollar bill.

Our Auntie Middo was the epitome of a proud American. She would arrive for visits always with a suitcase covered in travel stickers from most of the States and full of little nick-knacks for the kids. Most coveted were the silver dollars that I always gave to my brother to fill his hippopotamus. She also had the 'crazy' gene, particularly when behind the wheel of a car, and stories were always shared amongst the family of her hair raising exploits. Her son, Cousin

'Rosebud', a nickname for 'Robert,' was on the same nucleotide of the DNA molecule. He was an architect, Married Abigail Adams (also a much younger generation) and spent many of his office hours writing silly cards or creating some useful inventions like the paper box sheep (an origami paper box with eyes, nose and mouth on one end, a tail on the other and four hooves on the bottom. I used to love receiving his offerings in the mail!

Middo, short for Mildred, was the older and larger of the two sisters, my grandmother being the diametrically opposed version, somewhat like my brother and me, and the most beloved being of my life.

My 'Grannie' had lost the love of her life, my grandfather, the doctor, in her early thirties and was left with five children, four boys, and my mother as the only girl. He had died of pneumonia, and she never remarried, hiding his photo at the bottom of a dresser drawer, as it was too painful to view. She was the lead mare of our family, the keeper of its' principles, main object of adoration for all, and holder of all the family recipes; the love and nourishment that held us all together.

For many years she lived with my Uncle Doug, just blocks from our house. He had a tragic love story as well when my Aunt Audrey died during childbirth with her third child. Grannie then moved in to care for baby Patrick and his siblings Kelly and Hank. We were all very close as in the summers we joined Grannie at Roberts Creek with our cousins and in the school season would go swimming in Uncle Dougs' pool. Uncle Doug was my favorite uncle, and I had a huge crush both on him and my older cousin Hank. (In fact, Hank tried to introduce me to the 'facts of life,' but it wouldn't fit) Uncle Doug had been in Hollywood with Aunt Audrey (he was a radio host when radios were like T.V. and stars like Jack Benny, Bing Crosby, and Bob Hope were famous.) He bought the Commodore ballroom when he moved to Vancouver and owned it for years.

I remember he had a red convertible that was very flashy, and he took all the cousins for rides when he visited Roberts Creek. He also sent us gallons of colored popcorn and candy from the 'Nutty Club' beneath the Commodore. (This was long before anyone realized the link between sugar and hyperactivity! Or maybe not, and he was just enjoying the family game of entertainment via sibling abuse) My mother, being Grannies' daughter, was in charge of feeding, cleaning up after, and disciplining the cousins, (particularly Patrick and me)

The other store below was the 'Crack- a -Joke.' He took joy in placing 'whoopie cushions' for my mother to sit on by the pool or placing plastic dog vomit nearby, which my mother would push around with a Kleenex for a while until she realized it slid too easily and walked off in disgust. Audrey, his wife, had been a Reifel and very wealthy. I remember him coming to Roberts Creek on the 'Casa Mia', a yacht the size of a small ferry with slot machines and a spring dance floor. We had to climb up a 10' high ladder to get onto the deck. My ocelot coat, so distasteful to my future husband Kevin in university when we had disdain for each other in Spanish class, was a hand me down from the Reifels.

I remember Hank taking me into his room and playing Blue Tango, his parents' favorite song, and I carried those sadnesses of my Grannie and Uncle Doug through life.

I loved my Uncle Jack as well. He had lived with us until he gave up his bachelorhood in his forties and adopted my cousins Sandy and Susie, who moved in across the lane. It was wonderful having a cousin of the same age in Sandy, and I spent a lot of time with her there and at Roberts Creek in the summers.

Actually, the main crew at Roberts Creek became Patrick, Steve, Sandy, and I, and Patrick became the scapegoat for all our adventures.

If anyone got into trouble, it was Patrick. We were nasty, but we came by it honestly and always did it in good humour. He spent a lot of time in his room, which he shared with Grannie so she could keep a close watch on him. In fact, I remember him being sent off to camp Elphinstone for removal and reform, and returning with hilarious stories to make the adults feel guilty, like being locked in cupboards and having his meals served wrapped in newspaper. He didn't get sent off again, although his escapades continued. One of the best that he could truly own was at the beach, where my mother was in charge as a lifeguard. He flipped a rowboat and hid under the air space. My mother just about had a stroke, but in retrospect, when Patrick was discovered alive, everyone but her found it very funny.

My slightly older cousin Leslie would visit, and as we aged, she would come with a micky of gin and a pack of cigarettes hidden in her bag. If there was any scent of tobacco or alcohol detected by my mother, it was Patrick – poor guy, but we still love each other! Kelly rarely came as she was again slightly older and had a life, but I have one memory that became engraved in my psyche. She brought 'Mousie' a small stuffie the likeness of Minnie Mouse of Disney fame. 'Mousie' got wet (I'm not sure if Patrick was involved, but most likely) and Kelly put her darling little pet on the lampshade to dry. 'Mousie' was badly scorched, Kelly was devastated, and it tore at my heart enough to become one of those indelible memories.

For real animal pets, Uncle Dougs' family had Fritzie and Mitzie, who we visited on pool days. Fritzie was a dachshund who Grannie had fed to the point that his tummy dragged on the cement sidewalk and wore all the fur off his belly. Since he could hardly walk, he and Mitzie, an unremarkable grey cat (except for this), lay together in the front hallway close to the kitchen in each other's paws, Mitzi licking Fritzie all over as he also badly needed a bath. "P.U." we used to say, holding our noses when passing.

A magical memory from those years is the 'Tree House' at Roberts Creek, which was like a Swiss Family Robinson cross Garden House with its lattice shutters and proper staircase up to a deck at least 8 feet above the ground on top of a large stump. We would go up there and could see both directions along the beach. It was always the first place we visited on arrival when we first leapt out of the car at a full run.

Sandy, Leslie, and I, who were always housed together being close in age, would always scheme some exciting adventure (while planning how to blame it on Patrick) and would sneak out most nights. We bent low below the living room window sill as we watched the adults, oblivious in their after-dinner position, plumping our beds with pillows and escaping to join our friends across the properties. Our grandfathers, three doctors, had bought the strip of oceanfront when their children were small, so all of us had spent our lives of summers together. It was a magical time. All of the other parents were 'uncles' and aunts' or had nicknames like P.D. and Mame, and we all felt like we were related. It got a little incestuous at times as we aged together. I remember the girls would always outsmart the boys and we would lure the boys to the woods and dare them to see who could pull their pants down first for a little show and tell. Guess what? The boys always won, and we would laugh and throw pine cones at them. We would play games like kick the can, and our favorite, sardines in which you got to spend rather close time with a boy (what am I saying, all the boys you liked.)

We didn't mind sharing, and sometimes a particularly gorgeous and rare specimen (like cousin Nick from England) got to have his arm around Denny while holding my hand. We knew, but hey, maybe one of us would win out. I think; eventually, we had all kissed each other, had all been caught naked by parents, and all decided we needed to move on to someone not so closely related.

There were exciting events like when we were playing hide and seek and a black bear was hiding with us, or the time one came down to the beach, and my father picked up Leslie and I who were screaming so loudly that the poor bear jumped in the water and swam as fast as he could to a faraway point in the distance.

We went rowing and fishing every night after dinner, all in our different colored boats and ate fish for breakfast and dinner. We got a lot of bones stuck in our throats, particularly when the salmon was buried in Grannie's egg sauce (one of the recipes I never asked for). To this day, I can only do absolutely boneless fillets. We had campfires and camp songs and talent contests, none of which I won any awards in. I remember Patrick and I doing our classic dog trainer act every year; you know the one where the trainer says to 'sit,' and the dog lies down or 'rollover,' and the dog barks. (we thought it was funny). Denny and Leslie always sang with their beautiful voices, and I secretly practiced mine, singing Rogers and Hammerstein all winter, hoping to improve. Denny would have to put her fingers in her ears, so I didn't drag her off-key when I tried to sing along.

Such wonderful innocent times. There were no screens, just comic books, and 10 cent mixes, lots of outdoor activities and games and good clean, or sometimes dirty, meaning playing in the mud puddles after the rain, of course, fun.

When Uncle Doug eventually remarried, Grannie came to live with us and remained there until the end of her life. I could listen to her stories about when she was young forever. She and my grandfather had lived on Point Grey Road in Vancouver when the city was very young. They had a pony and a cow in their yard, and when they went to Roberts Creek in the summers, they traveled on the Union Steamship to the dock at the end of our beach where they would push the cow overboard, and she would swim to shore. Milk then was kept in a dirt cooler in the ground, clothes were boiled, and strangely all of the children were dressed in immaculate white sailor suits (I guess a fashion of the day) which had to be changed whenever they were dirty. I suppose that kept them busy all day. They went over the Burrard Bridge to Saint Paul's hospital in a horse and buggy, and everyone in the young city knew each other. In fact, Katy, who has rented our guesthouse at our ranch for years in the winter now, remembers stories of her grandfather then, who owned Worthington's drugstore, knew my grandfather and lived on Point Grey road. Those families' names are still on Vancouver institutions like Eatons, Woodwards, and Cunningham's drugstores, the family who bought out the Worthingtons. I often wished I had lived in those earlier times, and I'm sure I must have. I remember as I listened to Grannies' stories wanting to memorize every line on her face so I could never forget her.

In later years when I was in my Italian marriage and had my four young children in Surrey, my dad wound up moving his company nearby and bringing my Grannie to drop her off for the day, two or three times a week, where she would play with the children and entertain them with those wonderful stories.

One of the stories I loved was about Grannie's grandfather, a Crosby, another family name who was an MP in the Ottawa valley and lived in a stone farmhouse with window sills that were a foot wide. She told stories about feeding the milk to the chickens as the children drank the cream. They bathed in a large brass tub, which was heated and placed in the middle of the kitchen once a week for the children's' baths. He was a Presbyterian and built the town church nearby. There were stories about how it was a sin to go fishing on a Sunday, and the children hiding in the asparagus beds that were waist high and sneaking down to the river with their hand made fishing gear. All her life, Grannie talked about having a Presbyterian foot meaning a strict moral foundation.

I actually wrote my first book then, which was a children's book featuring my children and a year of their life. I did the illustrations and portraits of all the family members as they joined us for various activities. They were actually very good likenesses, and I still have a copy, the rest of which were given to the characters themselves.

Chapter Two

IN SEARCH OF THE MIRACULOUS

So on to my post-high school husband-hunting. A major contributing factor in my personal interpretation of life was the Rodgers and Hammerstein musicals, so beloved in the fifties and sixties. I had absorbed every aspect of those romantic renditions of true love, memorizing every lyric, every nuance in my ever voracious search for love. True love. The three musicals that stood out for me were the 'Sound of Music,' 'South Pacific,' and 'My Fair Lady.' True to my fidelity to higher guidance and therefore emerging abilities as a manifestor (Law of Attraction), I lived those three musicals in my understanding of life and the future choices of my husbands.

"Whenever the Good Lord closes a door, somewhere he opens a window." "Climb every mountain, ford every stream, follow every rainbow, till you find your dream. A dream that will take all the love you can give every day of your life for as long as you live."

So without forcing you to listen to countless words of wisdom, I will just add that I still include a singing day in the Equinisity retreats that I facilitate today with three or 4 hours of my favorite inspirational songs.

I must also mention, Leonard Cohen, John Denver, the Beatles, Bob Dylan and Neil Diamond among the other popular songwriters of the time as my life was shaped and moments were cemented in the truths of their words.

Again I have diverged from my post-high school education otherwise and secretly known to me as husband-hunting. At the time, I was armed with my Rodgers and Hammerstein view of life and was ready and willing to be directed to my husband and six children, animals, and an idyllic life in the beauty of nature (most closely resembling Christopher Plummer and The Sound of Music).

It appeared destiny had a bit of an educational side journey in the form of art school planned with my best friend Denny, who I had known forever through our historic family connection at Roberts Creek. Denny's mom Mame (Mary), had signed her up for the Royal Ballet School in London, England. In a last-minute decision in August of 1967, our mothers decided that I should join Mame and Denny in London at art school.

Now to find an art school very last minute. Well, true to my manifestation abilities, an art school was found close to Victoria station in London called the Heatherley's School of Fine Art. As the thread will bear witness to, it was another signpost in finding my husband of 32 years now, Kevin, whose family members had met there years previously.

So off I went in 1967 with Denny's parents Mame and P.D. (Peter Douglas), to see the World and look for my life as I had been directed in The Sound of Music. We found a flat at Baker Street station, and Denny was immersed in the long days at the Royal Ballet School, while I had a more liberal allotment of curriculum at Heatherley's. Immediately my radar tuned in on a charming American from California, there like me in search of diversion, but unlike me not yet resigned to find the love of his life. I was besotted and brought Otto to meet my English family now down to Denny and Mame as PD had to go home and make the money needed for our adventure. Otto was an immediate hit, and he allowed me to divert him until one evening when I announced I would be a virgin until I was married. That would be the end of our love affair. No amount of scotch on the rocks (my choice of drink designed to wow with my knowledge of an educated choice in alcohol) would convince me otherwise. I was sure that Otto would respect my higher ground and wait for me, but sadly that was one of life's heartbreaking lessons, and he actually formed a lifelong love relationship with Denny and a motherly version with Mame. Denny and I were sure it was Mame he really truly loved!

I returned home ahead of Denny, and when the news of their love surfaced, the result was a quick reassessment of the value of virginity, and I was off to the races. Although I still based my criteria on marital possibility, I realized that a little exploration could be fun and rewarding.

From Art school, it was decided that I should be a nurse. My family had a huge medical history from my grandfather, the Dr., grandmother and mother nurses, uncle Bob chief of surgery, and heart surgeon at Saint Paul's hospital in Vancouver (which posthumously resulted in the Gourlay wing of the hospital).

Being a nurse would be a good vocation to both find a husband and be a good mom. My cousin Sandy and I were both enrolled, but sadly both of us were more keen on the husband-hunting then the nursing.

Onward to university, and this is where the second thread connecting to my husband Kevin came in. No, wait, it was actually the fourth if you included that his uncle had met his aunt a generation ago at Heatherley's. Then the second thread was Kevin's Mother Margaret had grown up spending summers near Roberts creek and dated my uncle Bob the heart surgeon (before she went to England during the War and found Kevin's dad). The third was that after Kevin's parents immigrated to Canada when Kevin was 13, their best family friends were my dad's boss, and we attended Christmas parties with their young sons (the weird English boys). I digress, but the thread is interesting. It ever reminds me that my diligence to follow my truth and intuitive guidance from o. saving grace through all of my exploits.

So really, the fact that Kevin and I have fu. sharing a Spanish professor at UBC in 1969 / 70 fourth thread. However, the cloth was not yet ready fo and our memories were of a disheveled longhaired h. boy and an Ocelot coat clad airhead who aspired to a existence. Hence there was no opportunity for a deeper en. that time. I was off to find another piece of the puzzle as wa. so our final and fatal meeting resulting in marriage a renditio Brady bunch and grandchildren was not yet due.

At the beginning of my brief stint in university, I had wor a summer job drafting for my father at his engineering firm, and course, I was still living with my parents. I had amassed a smal fortune as far as I was concerned. My Doris Day philosophy told me it was, of course, to be spent and enjoyed. I invested in a Sunbeam Alpine sports car, not for any love of cars, but to get me around in style. My investment in love was in the form of a beautiful Palomino Arab Horse (Chako), as of yet untrained. I did not look at that as a hurdle or the fact that horses were expensive to maintain as I was certain that if I was called, those doors would open.

I enlisted my ever-patient father to help me with the 'how hard could it be?' job of horse training, purchased a book, and he followed the instructions while I became the rider. I quickly discounted the tried and true methods as taking far too long, the equipment like bits and saddles as being too cruel, and decided to teach Chako (short for Muchacho d'Oro, meaning 'boy of gold') English words and short phrases. "Chako walk," "Chako trot," "Let's go" It was fun to put Patrick or Steve on and halfway across the paddock push the let's go button! I also discounted the folklore about stallions being problematic and took him for forays along the roadways in search of the perfect grass in just a halter and lead rope.

Amazingly, he completely reflected my belief system and only occasionally did his manhood surface. He was always respectful of a stall with him, and a cute female ...ly, he would get an erection, and when ...it, he would settle down. Maybe I was ...me even...th a low libido, but I think stallions have horse w...

the ma...

luck...was a good time to move out on my own. That bee...rge old house in the residential Dunbar area of ...the flats where I decided to keep my horse and ...eater and Safeway with lots of leftover produce ...horse, not me). The problem was I was the last of ...o were sharing the house, and the only room left over ...alk-in closet. I was not deterred as freedom called, and ...d a window (unlike my closet outside the principal's ...as roomy enough, maybe 8 feet by 12 feet, and when I ..., being drawn to another possible life direction in interior ...I got creative with mirrors and graphics and produced a ...and white disguise that gave the illusion of doubling the space. ...ent hours enjoying the comforts of freedom, listening to my reel to reel tapes, and dreaming of my future home.

I mentioned my musical heroes briefly earlier, and I think my favorite of all time is Leonard Cohen. Over my life, there have been Leonard Cohen songs for every era, every twist and turn of my path. I think my very favorite is his last CD, 'You Want it Darker,' which reflected his ongoing rant at God. The words in 'If I Didn't Have Your Love' echoing my commitment and love:

If I Didn't Have Your Love
Leonard Cohen

If the sun would lose its light
And we lived an endless night
And there was nothing left
That you could feel

That's how it would be
My life would seem to me
If I didn't have your love
To make it real

If the stars were all unpinned
And a cold and bitter wind
Swallowed up the world
Without a trace
Oh well that's where I would be
What my life would seem to me
If I couldn't lift the veil
And see your face

And if no leaves were on the tree
And no water in the sea
And the break of day
Had nothing to reveal
That's how broken I would be
What my life would seem to me
If I didn't have your love
To make it real

There in the soft light of candles and the comfort of my solitude, I spent hours listening at the time to 'Suzanne' and 'Marianne,' not fully understanding the lyrics but soothed and healed by his golden voice. His music is one of the golden threads in my tapestry (I love you, Leonard-thank you for being a huge part of my life!)

Interestingly, when I look back at that time, it is not clear how I paid for all of my indulgences. I think I must have had some allowance from my dad (probably as I had agreed to finally go to university, and he was able to issue a huge sigh of relief). I do remember it not being quite enough to support all of my investments in the life I knew I deserved. Very quickly, the gas and car insurance, the horse board, and my board was draining my meager savings. Inspiration came when I noticed we had a huge yard with long grass, and we were a short drive from the flats and Chako's stable. I enlisted my roomie friend Anthea as the driver and sat in the back seat with Chako on a lunge line calling "Chako walk," "Chako trot," as he eagerly followed behind. I had, of course, prepared the fenceless yard with a string line at eye level in lieu of a more expensive solution. I had been told that horses would not challenge that.

I experienced a few days of bliss sleeping in my pasture with my horse, waking up my roomies to the sound of hooves as he chased me around the yard, and depleting their stores of breakfast cereal for his grain.

It all ended rather abruptly when one day we got a call about our horse being over at the movie theater attracting a crowd, and the bylaw officer wanting to speak with his owner. Well, according to the officer, if I had had an elephant in the yard, there were no bylaws in place, but the horse had actually been ruled out with urban planning. What was I to do but find him a country home, and move on? I still held the vision of my perfect life with horses, which would eventually manifest in my life of the last 20 years with my Herd and Epona my Chako reincarnation.

Husband number one was to be found not at university but in the form of my German landlord suspiciously resembling Christopher Plummer in the Sound of Music. That the resemblance was purely superficial became apparent and resulted in an abortion when he told me he did not want children. A divorce came shortly thereafter courtesy of PD Denny's lawyer father and our long-term family friend.

I moved back home to recoup, rest, and meditate. One evening lying on my childhood bed, I had my first God experience in the form of total immersion, as Joel Goldberg had shared in 'Beyond Words and Thoughts.' It was beyond words and thoughts. There was a loud buzzing, and I lifted above my body, up up, to become one with God. Floating in a sea of LOVE. I absorbed a profound understanding that has remained my foundation since that moment. It was all okay, we were one with God, and God was all that is! I then experienced birth and re-entered my body, knowing that I had come to do something for God, and that was the most important mission of my life.

Onward and upward, yet again, I moved back in the direction of art, studying with a charming senior instructor from the Vancouver School of Art, and well-known artist Peter Aspell. I began to paint with excitement and determination producing a varied body of work from portraiture to highly textural creations using tree branches, mirrors, resins, fur and nails (not all on one piece)

After a brief stint in nude modeling to pay for my art courses, I attracted an agent in the form of a mature student skilled in business. I garnered a centerfold in the leisure section of the Vancouver Sun newspaper (not for nude modeling but for my artwork) and a one-woman show at the Vancouver Home Show. I was off to the races, sold some pieces and experienced a one night career ending in my about to be Italian husband (who resembled Ezio Pinza in South Pacific) sweeping me off my feet behind the curtain while helping me hang my one hundred pound nail sculpture (I'm amazed he had the stamina left to breathe, let alone take my breath away)! A conveniently produced diamond engagement ring was presented after that night, and we were off to a seven-year marriage and four children. One little hitch was his divorce from his first wife aided by an ex-boyfriend of mine who happened to owe me and was also a divorce lawyer.

Those seven years were pretty idyllic as we wound up on a lovely acreage in Surrey as a result of his introducing espresso coffee to Vancouver. It was great until it wasn't. I didn't see much of him as he was on night shift in the city, and I was growing babies and vegetables, baking bread and making my own pasta and creating more art in the barefoot and pregnant attire. I had a brief taste of my country life with animals and children. It actually ended abruptly when I discovered he was sexually inappropriate with my two eldest daughters then three and five years old.

With a little help from above, he conveniently was called away for a weekend. I called my father, (who was the epitome of a perfect dad), and the one day movers and left him his things and a note threatening jail if we ever saw him again. I'll never forget the 'last supper' when I was sneakily packing while pretending to sort cupboards and drawers as he followed behind, checking over my shoulder and rearranging everything. It was truly stressful as I was terrified of him by then and afraid that the children would drop one too many peas on the floor or smear too much tomato paste on their faces.

Of course, the phone kept ringing, asking if we were all right, and I kept thanking them for calling coming up with more and more veiled replies. Then there was the final conversation after the children were safely tucked in bed in which he told me he thought if we would ever part, I should leave him the children. I answered with my most "we'll see" voice while sending my spiritual armies to take him down. When I heard him outside sorting through the barn (I was sure it was for his rifle), I called sweetly for him to come to bed and seduced him one more time for my country in an attempt to settle him down. Yuck! He fell asleep satisfied and was gone in the morning. We never saw him again, and I thank my lucky stars that my children, all four, were under the age of six and hopefully would all recover.

That was a period of deep reflection and spiritual growth for me, a single mom who had lost all desire to ever have another man in my life. I couldn't even look at a man and, in fact, was alone on my path of self-development for nearly seven years. The country home was sold, and we rented in a child-friendly neighborhood close to a school and my parents, my dad becoming a surrogate father to my brood.

Being a mother of four children, the oldest of which was six, was all-consuming in the daytime, but when they were all bathed, read their stories and tucked into bed, the evening stretched ahead of me, a gift in self-introspection and education. I examined my life thus far, my reason for being. How had I've been so misled? But then again as my pediatrician (again an extended family member from Roberts Creek, Dr. Campbell, whose family lines had extended from my grandfather's best friend Dr. Covernton's wife) said: "You have four beautiful children and they are all healthy. " It was true, and I was only two short of my childhood prediction that I would have six kids and live in a country home, surrounded by animals. The only thing missing was someone to take on the role of my Prince Charming.

I began to explore my spiritual callings from learning to meditate to immersing myself in my all-time favorite authors like Kahlil Gibran and 'The Prophet,' which I had memorized years before during my sojourn in the cupboards of my life. 'The Prophet' on love, marriage, work, houses, joy, and sorrow.

Here are some excerpts from Kahlil Gibran -The Prophet

Love

When love beckons to you, follow him,
Though his ways are hard and steep.
And when his wings enfold you yield to him,
Though the sword hidden among his pinions may wound you.
And when he speaks to you believe in him,
Though his voice may shatter your dreams
As the north wind lays waste the garden.

When you love, you should not say, "God is in my heart," but rather,
"I am in the heart of God."
And think not you can direct the course of love, for love, if it finds you worthy, directs your course.
Joy and Sorrow
And the selfsame well from which your laughter rises was oftentimes filled with your tears.

Giving:

And there are those who have little and give it all.
These are the believers in life and the bounty of life, and their coffer is never empty.
There are those who give with joy, and that joy is their reward.
And there are those who give with pain, and that pain is their baptism.
And there are those who give and know not pain in giving, nor do they seek joy, nor give with mindfulness of virtue;
They give as in yonder valley the myrtle breathes its fragrance into space.
Through the hands of such as these, God speaks, and from behind their eyes, He smiles upon the earth.

Chapter Three

THE SEARCH CONTINUES

I discovered Dr. M. Scott Peck and 'The Road Less Traveled' and had my second God experience while reading his section on Grace, and being given my second microchip of awareness that changed everything. God was not a kindly father figure that was somewhere above us in the sky. He was 'Us' like a Rhizome from which we individual flowers emerged, grew, and died, feeding the rhizome with our experiences. I went into a transcendent state of realization that lasted for at least a day, dancing around the house and singing the words to 'Amazing Grace.'

Amazing Grace

T'was Grace that taught my heart to fear.
And Grace, my fears relieved.
How precious did that Grace appear
The hour I first believed.

When we've been here ten thousand years
Bright shining as the sun.
We've no less days to sing God's praise
Than when we've first begun.

I'd then went on to Thomas Merton:

Thomas Merton, New Seeds of Contemplation:

"A tree gives glory to God by being a tree. For in being what God means it to be, it is obeying Him it "consents' so to speak, to His creative love. It is expressing an idea which is in God and which is not distinct from the essence of God, and therefore a tree imitates God by being a Tree.

Therefore each particular being in its individuality, its concrete nature, and entity, with all its own characteristics and its private qualities and its own inviolable identity, gives glory to God by being precisely what He wants it to be here and now, in the circumstances ordained for it by His love and His infinite Art.

The special clumsy beauty of this particular colt on this April day in this field under these clouds is a holiness consecrated to God by His own creative wisdom, and it declares the glory of God."

So many of the books I digested at that time of my life (early thirties) became my references at the retreats some 20 years later. I sped read 'A Course in Miracles'

From 'A Course in Miracles:

This is a course in miracles. It is a required course. Only the time you take it is voluntary. Free will does not mean that you can establish the curriculum. It means only that you can elect what you want to take at a given time. The course does not aim at teaching the meaning of love, for that is beyond what can be taught. It does aim, however, at removing the blocks to the awareness of love's presence, which is your natural inheritance. The opposite of love is fear, but what is all-encompassing can have no opposite.

This course can, therefore, be summed up very simply in this way:

>Nothing real can be threatened.
>Nothing unreal exists.
>Herein lies the peace of God.

Then onto 'Beyond Words and Thoughts' by Joel S. Goldsmith.

"If there were a God in the human world, rape, arson, murder, wars, dens of iniquity, drug addiction and all other afflictions of humankind would be impossible" … I have never doubted that there is a God, but now at least I know there is no God in 'this world' "

He goes on to say, "Where is God? What is God? How can we have an experience of God?" the answer is meditation, surrender.

As in The Sound of Music, the council is you have to look for your life, and from the bible, "ask, and the doors will be opened."

"Climb Every Mountain" - From Sound of Music

>Climb every mountain
>Search high and low
>Follow every byway
>Every path you know
>Climb every mountain
>Ford every stream
>Follow every rainbow
>Till you find your dream
>A dream that will need

All the love you can give
Every day of your life
For as long as you live.

As I continued to ask and search for answers, my connection to higher consciousness grew. I began to feel the presence. I would get a tingling in my Third Eye area and know I was being called to tune in, and as I began to tune in more and more, my healing powers began to develop. I started to feel called to do hands-on healing, and over that time and on into my years in Gibsons, I experienced several miraculous healings. It was actually embarrassing as I felt bound to follow my guidance sometimes amid a skeptical audience. A couple of particularly embarrassing situations come to mind.

One was early in my career when my childhood friend Patti-Anne had a baby diagnosed with meningitis. Her mother had died of it, and I was called. What was I to do? I went to the hospital and offered my services. The baby was in isolation, surrounded by monitors and doctors all in full isolation garb. I was quickly capped and gowned, sterilized and brought to the bedside. No pressure! I put my hands on the baby and went to the place where Joel talked about 'The experience of God' and waited for confirmation that God was, in fact, there. The monitors did a big blip, Patti-Anne pushed her nose against the glass, and the baby responded - it was a miracle! He turned for the better, and Patti-Anne believes I saved his life. I have no idea what the doctors thought.

Then my most embarrassing calling was years after in Gibsons when Kevin and I were finally together and working with a big developer on a very expensive project. It was a lunch meeting, and 12 of us were gathered around the table. The developer was sitting to my immediate left, and we began talking about going to the site for a walk around. He began complaining about his right knee, which was right next to my hand. There it was again, a tingling in my third eye, Oh no! I had to do it as it was my promise to God. I gave him a brief outline of what I was going to do, the table fell silent, and I went to my experience place as best I could with the whole table patiently watching. Sheesh! Well, 5 minutes later, it felt done, and I took my hand away to a barrage of 'she had her hand on my knee' jokes. The developer got up, walked around the room, and said he was ready to go on site. It never bothered him again, and he put the money up for the project. I should've asked for a cut!

My healing career carried on well into my years in Gibsons. I had maintained health for my entire family from Martin's chronic bronchitis to Cara's chronic ear infections to Kevin's problematic back (he was bedridden for a week prior to our wedding, and then never again in 32 years). I do so totally believe that most of health is a benefit of a healthy diet, clean water, good self-care (exercise, sleep, etc.), and positive thoughts.

I also believe in energy work. The precepts of modern medicine, A.K.A. vaccines and pharmaceuticals, have gotten way out of hand. They're the leading cause of death in North America. Recent trends in the medical are toward functional medicine which looks at the cause and examines overall lifestyle rather than putting a band-aid on the evidence of dysfunction, like the paramount statistics of inflammatory diseases which are everything from cancer, diabetes, heart disease, autoimmune issues, and mental disorders to name the biggest. Our environment is increasingly toxic as well with poisons, GMO, electromagnetic radiation, and polluted drinking water. Sorry, this was supposed to be a light and entertaining romp. Back to that.

So my healing abilities developed during that time alone. I had plenty of time to self-improve, and I was ready to get back to my 'Looking for Love' phase of life. Just a footnote: when I visualized my life as two books, the first was 'Looking for Love' and the second 'Naked into Truth.'

Fran, Peter, Alex and Laurie in Richmond before meeting Kevin

Peter and his Nono

Grannie with Alex

Chapter Four

THE SEARCH BEARS FRUIT (OR; UNDER THE APPLE TREE)

After a brief reawakening with a darling family friend from my extended family at Roberts Creek, which became the 'Road Not Taken,' it was time for Kevin to resurface in my life. He, in the meantime, had married, fathered two children, and kissed his wife goodbye when she ran off with a famous opera singer. His mother, called by spirit to further our cause, had found Kevin's wife a job at the Vancouver Opera and effectively introduced her to her star, ending her son's marriage. Well, of course, there was a lot of upheaval and soul searching on Kevin's part, but again with a little help from our higher up friends, he was left with the house and acreage and full custody of his children. An unbelievable precedent and the perfect setting for our happily ever after story (He was Rex Harrison in 'My Fair Lady' and a true gentleman). Not only that, but he had the perfect house for our six children having become an architect in his years after Spanish class.

It was definitely the Brady bunch. I had found him again when he surfaced to take Denny - now living permanently at our summer place in Roberts' Creek - out for a date, again courtesy of his mother who was well connected in the theater and knew Denny was a newly single dancer. I, of course, felt called to go and check out Denny's date as she and I were neck and neck for who can remain single the longest. I thought him perfectly acceptable, and she had no interest, so the door or window that the Good Lord opens, opened.

He arrived at the beach the next day in his little red sports car, a Triumph TR4 to pick up his ever-elusive wallet, which he had left behind the night before. In 32 years of marriage, Kevin regularly loses (actually, only briefly misplaces) his wallet, glasses, keys, etc. and always manages to re-manifest them. I think it's Angelic help. Anyway, conveniently, he found Denny and I sunbathing at the beach, me with my 'Course in Miracles' book open and at the ready. He very politely asked if either of us would like to join him at a meditation group that night. Denny declined, and I almost jumped in his lap, so that became our first date. I had never been to a meditation group, and I was over the moon as I had only meditated in my own company and was thrilled at some coaching.

That evening was the beginning of a divinely guided two weeks, a whirlwind ending in a marriage proposal and the end of my search for true love. I will digress to the details. I was so rusty from my years of conviction that I did not want a man, didn't even want to look at a man, but from that first date, destiny took charge. After our meditation group, somehow Kevin's little red TR4 gave a gasp and decided to take a long nap at the bottom of our trail. We went inside to give it a rest and talked and talked until well after midnight. Since I was still under the wing of my mother, who was in her cottage on the property only a few yards away, I suggested he take my car home and come back and deal with his the next day. A day of tinkering ensued, and by the next afternoon, all of the neighbors were wandering over having heard the rumors that Lizzy had a date. There were six children involved, and by dinnertime, we had a large meal, some more talk, and then Kevin decided to give the car one final try. Well, it bunny hopped up the trail, gaining full momentum when it reached the top, probably pleased in its role in the longest date ever.

More nights of talks followed, but a week later, my buried libido had not noticed a whiff of pheromone and was wondering why this guy was doing so much talking. My girlfriend Patti-Anne of the miracle healing fame who had come to visit, when queried, suggested he was cute, and maybe he liked me. That evening he decided his move. Putting his arms around me in an affectionate, good night hug, he placed his armpit directly over my nose while giving me a light kiss on the top of my head. What just happened!? That was the signal my higher-self needed to remind me of our earlier agreement to spend this life together.

I couldn't sleep, tossing and turning, desire swirling; I was a weakened shadow of myself by the next day. That was the day we had planned to overnight at Rainbow Lake, a local camping spot, with all six children. During the day, I had to withstand the breath-stopping handhold, the casual arm around me, and a buff body in a swimsuit. I was done! That evening when we had zipped the kids into their tents, we tore each other's clothes off stumbling to the mummy bag under the apple tree, which I had placed far away in resolve to not do what we were doing. You can't get much closer than in a mummy bag, and Kevin wrote in his journal the next day 'Liz + stars + skunk + squished + rocks + moon + apple tree.' It happened to be 1987, August 16, the Global Convergence, where people all over the globe were meeting in peace, love, and harmony to herald in a new age for the Earth. I should also mention that at one point, he kissed my fingertips, and a neon sign flashed in my brain, "He's the one." The next morning Kevin gave me his family ring, and we were one.

It was a few days later that we had time alone, a date night at Kevin's home with all of the kids farmed out. I had not yet seen his house, and with Bob Dylan playing 'Lay Lady Lay,' I explored the roomy three-floor post and beam palace with a view over the treetops to the ocean. It had thirteen acres of pasture and trees, a cottage that became our teen home, and I was sold. In fact, I had sold my home in Surrey and was looking for a new place. Kevin's place was up for sale. It was close to empty as his ex had removed all of the furniture, and I had lots of furniture and no house – a perfect fit. "I love it, I'll buy it," I said, looking over the treetops to the ocean. Being my prince, he put his arms around me and said, "You don't have to… Marry me" (which I'm sure he regretted sometime later when he had to pay the bills).

There was considerable planning involved for our two families to happily co-habit and blend smoothly into the new to be shared space. I had the furniture, Kevin the house, but there were the logistics of space and what and who would fit where. In spite of the house being mostly devoid of furniture, there were 'personal possessions' that were near and dear to Kevin that soon raised their profiles and a new-to-be-discovered aspect of Kevin that has been a topic of disagreement for 32 years now.

Kevin is a hoarder. Liz is a minimalist. For example, the bedroom closet space. Kevin's side was spilling over with towering heaps of sweaters. Apparently, the ex-mother in law was a knitter, and he was blessed with countless gifts. They were all wool, beautifully executed, but sadly over the marriage had fallen prey as a favorite food for moths. One by one, I would examine them to see if any could be saved, and my criterion was if there were one or two holes that possibly would not be noticed they stayed. If they more closely resembled swiss cheese, they were designated for the dump. The black bags would fill, and not just with sweaters but socks, ripped shredded or oil-covered, much-used work ware, contents of drawers, like the bathroom drawer with used Band-Aids, Q-tips, empty vials of whatever and even dried peanut butter sandwiches. Every weekend visit was a new chance to sort, clean, and prepare. Also, every weekend resulted in a painful extraction of the familiar from Kevin. He would be sent to the dump with a pick up full of black bags, and after an inordinately long delivery trip, Kevin would re-appear with half the bags all carefully re-arranged. I might have become more wary had I discovered the old love letters from high school that were well hidden in the bowels of the basement.

It was apparent we really loved each other as there was only one incident when we had a stand-off and when I said, "It's them or me," Kevin picked me up and put me in a black bag! More on this amusing incident in a later story.

By Christmas break, we moved in and were married the following March with all six children lined up at the front of the little white United Church that Kevin had designed. It was very Sound of Music with John Denver's "You fill up my senses like a night in the forest, like the mountains in springtime, like a walk in the rain… Come let me love you, let me give my life to you. Let me drown in your senses; let me die in your arms…" sealing our vows.

We honeymooned in Kevin's four-poster with jeweled star-lights above and a big balcony overlooking the surrounding trees. Those years of parenthood and new love became a magical time with young children, big Christmases, and games like murder in the dark, which was our favorite as we could hide in the dark and make out while the game was being played. There were baseball, rugby, and soccer games, weekend hikes up the mountain, skiing, parties, and family gatherings.

Throughout, Kevin and I maintained our Friday night date night, holing up in our bedroom, which was out of bounds to all but Peter, who insisted on bursting in, sneaking in, or lying outside the door complaining he was dying of some malady. We set booby traps for him and carried on. We were often completely oblivious when in each other's company.

Fran, Alex and Laurie, in Robert's Creek at Kevin Meeting Time

Kevin staking claim to Liz

The House in Gibsons

Denny and Liz at Robert's Creek when Kevin instigated the Meditation Night

Kevin's Parents Desmond and Margaret

Cara, Martin and Peter when the family was formed

Liz, Kevin and Uncle Billy at the wedding

Christmas morning in Gibsons

Kevin and Liz enjoying a Peter break-in

Liz and Kevin when they met

Chapter Five

LIFE IN GIBSONS AND THE BRADY BUNCH

I think the largest gaffe was when our teenage daughters hosted a party in their 'teen house.' We had warned them "only by invitation," "keep it small," etc. But by the time we had snacked, videoed, and enjoyed our evening, we noticed a bit of a din and wandered out on the porch to have a look. Oh, my God! The yard was full of vehicles, all 'dad's cars.' There were bodies spilling and jumping out windows, and we decided we should call our local law enforcement to come and help.

When they arrived, we heard they had knocked on the door twice already (it was a long way from our bedroom), and they really couldn't do much, so they left. It was at this point that Kevin had a brainstorm, and pulled out his beat-up old land rover, the farm vehicle, jimmied a spot in between parked cars, then walked into the party and announced: "I'm going to ram any vehicles in my way that are still here in 2 minutes". He then proceeded outside where he revved the land rovers engine as loud as he could. The partygoers left in their daddy's cars in a big hurry, mumbling about the crazy man and leaving devastation in their wake. The girls had quite the job cleaning feathers from the quilts and pillows off the sticky beer-soaked carpets, filling holes in drywall with plaster, and learning new carpentry skills. After that, we bugged the teen house with hidden baby monitors and had a listen whenever our intuition alerted us (like the time boys were asked to go home and they snuck back in to find Kevin waiting on the porch). We told all of the kids that we could see right through them, and they really believed it until they found the monitors.

Another memory comes to light of Peter, of course, who had a friend sleepover but they didn't sleep. When we went to check on them before bed, we found their shoes by the door, and the boys were nowhere in sight. We couldn't figure out how they could be any great distance away, walking around in their socks. Of course, we didn't consider the Rollerblades, which got them down a mile of gravel road (don't ask me how that was possible) to the big town of Gibsons and eventually to Peter's friend Tim's house where they overnighted. It wasn't until they didn't return by noon the next day that I decided to call Tim's mother, and she discovered them sleeping in Tim's room in the basement. It turns out they were bored and hungry and thought it would be a great idea to Rollerblade down to Tim Horton's. Once they had finished stuffing their faces with doughnuts, they thought it would be much easier to Rollerblade down the hill to Tim's house, rather than skate hiking back up the hill to ours.

I was thinking I could write a whole chapter on Peter as he was always very creative and always had a new adventure up his sleeve (He actually went through a phase of wanting to be a magician which was utterly hilarious). For now, however, I will stick to some of the highlights like the time when he had another younger friend for a sleepover. Kevin and I heard the sound of gravel under car tires as they rolled the car down the driveway towards the barn at just after midnight. We were in hot pursuit as both boys were not yet licensed (Peter had just received his learner's permit), and we finally caught up with them in lower Gibsons. We were driving slowly through the pouring rain and began to flash our lights at them. They apparently spooked, and the chase was on along a windy high road above the ocean. We slowed right down as they shot off into the distance only to find them around the next bend, on top of the concrete barrier at the side of the road with the front wheels of our Mazda MPV Minivan teetering over the edge and a fifty-foot drop down a cliff and into the ocean just six feet to the right of the vehicle. I thanked my angels, guides, and whoever was listening and fessed up to the police officer (after lying that I was the driver) that a squirrel had run across my path, and I had swerved to avoid him. Peter owed us for some time after that, and if I recall correctly, he had to sell his computer to pay for the damages to the van, which was akin to a life sentence in his eyes at the time.

This reminds me of another time when he was around 11 or 12 years old, and he saw me writing a cheque and curiously asked what it was. I gave a simple response that you can use a cheque to get money from the bank and give it to someone else. Little did I know that Peter had interpreted this to be a magical way to somehow dupe the banks into giving you free money. Later that week, I received a phone call from the bank that he had tried to cash a cheque for $100 filled out and allegedly signed by me in pencil. They had politely refused the cheque because it had been written in pencil, not ink, which saw him returning the next day having traced over the pencil with a Bic pen. At this point, Kevin and I decided that he definitely needed to be diverted from a life of crime, so we called our friend Ed, a sergeant in the R.C.M.P. who handcuffed Peter, gave him a tour of the local jail cell and briefly locked him in there. Thank God that was the last of Peter's criminal career.

Other stories involve a little less life-altering danger like the time when he snuck into our bedroom in search of hidden treasure and discovered a bag filled with adult materials hidden at the back of our dresser drawers, in fact, hidden behind the drawer if I remember correctly, and then produced it at the perfect moment when friends were visiting, and he could dangle them in front of his mortified mother. Enough!

Just a few other kid stories as I got a little carried away with Peter's adventures. A time that stands out in memory involving both of the brothers (Peter and Martin) was when they decided they were going to build flying machines to get to school (those creative genes from both Kevin and me). It was a big project, Martin having conceived the idea to build a hang-glider out of alder trees and bedsheets and Peter (and even for a short while, Cara) following suit. Martin was the only one who had purchased a second-hand bed sheet large enough to cover his frame, so Peter immediately volunteered to become the test pilot. On launch day the boys were so very sure that their design would actually work (they were only about eight years old at the time) and we all lined up to watch as Kevin, Martin and Martin's schoolyard friend ran Peter and craft down 20 feet of hill and threw Peter off the eight-foot embankment at the end of our back yard. Crash landing into the grass below, Peter swore that he actually got lift, and unbelievably this gave Martin the courage to make an attempt himself. Alas, three weeks of effort ended in a battered glider and no more than twelve feet of induced flight, but I think the laughter was well worth it.

Unlike Peter, Martin was not usually one to test limits. His claim to fame was that he became my father Len's favorite when he discovered a passion for golf (a passion my golf-obsessed father had unsuccessfully tried to pass on to Peter for most of his early years). Martin lived golf all throughout high school and was the only protégé in the family. He was really good and, for years, wanted to become a professional much to the delight of Len.

The only sketchy thing I remember about Martin was 'the beer incident.' He had been given a new camel pack and ingeniously sat under the stairs of the sundeck and filled his pack with four bottles of my bartered homebrew, which he then smuggled into a high school party. I only discovered the empty bottles weeks later.

Another joint effort, I think, but I'm still not sure, was the 'flying saucer attack.' This again was not discovered for some time as it's not often that anyone examines the roof of their minivan. Mysterious saucer-shaped indentations along one edge of the roof appeared in assorted sizes. We couldn't, and in fact, didn't figure it out for years, just claimed it on our comprehensive insurance with a shrug. The boys fessed up many years later that having been parked in the perfect spot adjacent to our trampoline, they decided it would be a great launching platform in an attempt to obtain the ultimate bounce. A short while later, they graduated to jumping off the overhanging balcony, and the mysterious new saucer impressions ceased. You'd think they would have noticed their footprints in the roof and stopped!

Cara, along with Laurie, were our horse girls over the years learning jumping and eventing and competing at the local shows. Cara was fearless. On her first ride with older sister Laurie, they went for a trail ride down our long gravel road connecting us to the highway below. I was standing in the carport upon their full-out-gallop return. Cara, instead of being in position on Tao's back, was clinging upside down underneath his neck. She let go into my arms and, after a brief moment of shock, began giggling with delight and could hardly wait for the next ride.

Franny and Alex co-habited in the teen house, previously mentioned in the wild party story and our bugging protocol. They loved and fought and got into many adventures together, several that I'm sure we never heard about.

I really do enjoy all of my children's individuality. Today they have all flowered, most have borne fruit, and remain uniquely individual in their callings. Starting from the oldest, Alex is an entrepreneur, having developed a spice blend and quick and easy food business called 'Made with Love.' It is all organic and uses only biodegradable plastics or glass bottles with metal caps. She has also turned her half-acre property into an Air BnB paradise with an urban farm, a treehouse (more like tree cabin), a Teepee, and a cottage all available to experience (chickens and bunnies visit all guests). She is mother to four of my grandchildren, who have held the ranch together for the duration of the retreats.

Francesca is still living on the Sunshine Coast (after a very short stint in Vancouver), which is a renowned artisan community, and both she and grandson Gabriel, now 17, are involved in the Theatre and dance community that is the heart of the populace there.

Gabriel is a brilliant actor and tap dancer, and when you see him light up a stage, you know he has found his calling. He can recite Shakespeare on prompting while in commanding full character (I'm sure he's channeling) and is fulfilling my aspirations to be a theatrical star. Franny is a nature lover, healer, and definitely embracing her ancestral gatherer instincts, having a wealth of knowledge about harvesting food and medicine from nature.

Laurie is a beautiful, talented, and committed mother to her two girls Chiara and Mariah, who both inherited my artistic talents. Mariah commits the most amazing paintings to canvas, and Chiara, who is a photo-realist in her artistic talent, is diverted at the moment to fingernails. The only problem is that Laurie is incredibly allergic to animals, which has unfortunately relegated our visits to her-house-only. This means I only see them a few times a year because they live in Surrey, which is a significant distance from the ranch. Laurie of all my children inherited my all-consuming love of God. Having said that, Franny is a close second.

Peter is a computer expert, and my website and graphic designer guru. He has created all of my websites and is also my book editor, copy editor, photo editor and designer of the covers for several of the books. He's also the one who helps me to put together all of the manuscripts. Sometimes I will look at my computer screen and see it doing things that I haven't asked it to, and Peter will call me up on Skype or remote access into my computer and help me solve the problem. He's also designing his own video games and will take any opportunity to expound about his ideas even if I don't often understand a word he's saying. Regardless, it's good to see his passion.

Martin and Cara are in Ontario, the furthest away from all of us, but have always been the least concern. Like their father, they are studious, having both graduated with degrees, an MBA in business for Martin, and Law for Cara. They are both self-reliant, kind, and compassionate first-rate citizens. They are the most recent to produce, Cara having two young girls and Martin a toddler (also a girl) who are beginning to morph into their intended selves.

Chapter Six:

WORK IS LOVE MADE VISIBLE

Throughout the Gibsons years, there were lovely stories of grandparents and family gatherings. Kevin's parents lived in Gibsons as well, so we had many family Sunday dinners and occasional dinners at their house sans kids, which were a welcome break at the time.

They were so supportive of our careers, successes, and needs, and during that time, we were both building our careers, Kevin's in architecture and mine in Wildlife Art. We became notables in our community as Kevin had designed most of the prominent buildings in town, and I acquired an investor for my fine art prints and an agent to market them. My agent was very good at PR, and since I was affiliated with any charity who asked, from wildlife groups to Cystic Fibrosis, I was in the news weekly.

My art business grew to include doing my own printing in Vancouver at an office in Gastown, which I commuted to several days a week and many art shows. I was selling my originals for very good money and still doing mom duty. I really got good at meditation and could do it in the middle of a busy kitchen, producing a barrier of cotton wool-like cloud material and just disappearing. I was always creating something from a new painting or project to a new recipe for healthy kid snacks.

It was during this time that I invented "Better Butter," a half and half blend of healthy oil and butter that remained spreadable after being refrigerated (it has recently resurfaced in an adulterated form with additives 20 years later). I then embarked on a healthy Rice Crispie square as I was worried Kevin would get hyperglycemic the way he downed them. I crumbled up rice cakes, mixed them with peanut butter and honey, and everyone loved them. That was the beginning of my Food line as I made an appointment with the rice cake manufacturer (who also wound up buying several of my original paintings) to sell him my idea. That idea resulted in flavored rice cakes appearing on store shelves a year or so later. Well, during that conversation he told me about 'the culls', all of the broken rice cakes, that needed to find a use and I went back to my laboratory, blending 'Lite powder' an instant thickening agent that was perfect for 'lite' salad dressings, jams or other assorted 'lite' products from mayo to soups, brownies, whatever.

'Tree of Life' admins hellied up for a full course lunch sampling and were sold. I was offered 2% of the profits and an invitation to develop the product line. I even illustrated the booklet. It looked very promising until my liaison from 'Tree of Life' got sick, and an incompetent woman took over, taking lids off all my shipped samples and somehow having an inability to re-create them via the simple and detailed instructions. I was left with the choice of going down to Florida and developing the line personally or staying with my family and art (also the animals which had begun to mount with horses, dogs, and cats – the beginnings of my next career). It had been fun while it lasted, but it was an easy choice.

I began to barter with my bigger pieces of art and became good at trading for washing machines, rugs, boats, and even new Mazda vans. Everyone knew me, and I had regular visitors to my studio who often bought groups of 3 paintings just to round off the figure.

I had many images with B.C. Wildlife Artist of the Year, Ducks Unlimited, and even a Habitat Canada Conservation stamp. I was also invited to submit a design for the Canadian two-dollar coin and given a tour of the mint in Ottawa. I had investors who bought and donated prints to various conservation groups for a big tax write off. I became the fundraiser of choice as many of the heads of conservation groups like Western Canada Wilderness, B.C Spaces, World Wildlife Fund, and Canadian Parks and Wilderness were friends in our little offbeat community of the Sunshine Coast.

We were often offered rafting trips in Northern B.C. and Alaska. I would produce the flagship painting, which was presented to dignitaries throughout the world from our Premiere to Al Gore and Prince Phillip, basically encouraging them to protect areas for wilderness and advertise with an original of the area on their wall.

There were many articles in the paper and magazines, and one day an investor came forward to fund my new inspiration 'Silent Whispers Publishing.' I would publish my own prints and hire an agent, and we even went so far as to start our own non-profit 'Earth Wild International,' but it didn't go very well being not at all hands-off and actually compromising our investor donations.

Somewhere around this time, Kevin actually took a year off architecture and managed my growing career. We started making some real money, and I moved to an office in Yaletown, a funky area of Vancouver where I commuted a few days a week. Printing was fun! I found a big printing house and convinced them to expand into Limited Edition Prints. The process was fascinating, and I had to educate myself from the ground up. The resulting prints were beautiful, and we won awards for them. Bowne (the printing house) had several floors in a shiny glass high-rise with board rooms thirty feet long (complete with numerous phones, bar fridges, and an enormous conference table (aka the T.V. show 'Suits'). I became the office darling being young and attractive, clad in short skirts and silk suits. Jacques was my liaison and gave me full reign in the boardroom to sign and package my prints. I took up residence for a few days when a print run came off the press. Jacques would take me for lunch in the revolving tower nearby, ordering wine and smoked salmon, lobster bisque, the works. He was funny, and we had a great time whiling away two hours per lunch (tough job we had).

Another fun date was always with my investor Ken who was adorable (He had a kind of cowboy look when compared to Jacques in his Armani suits). Ken would come to inspect product, lunch me as well and even visit us in Gibsons for an update every once in a while. I remember one memorable dinner we had at the top of some other high-rise restaurant. I was paying him back for some of his investment, and we were looking at a contract for our continuing relationship. Always a little out of body, I held the contract over a candle to have a better look, and it went up in flames! Oh my, God, did we laugh!

My big business career was so much fun. I was offered another place on Howe Street. A P.R. lady took me on and offered me a section of her office to keep prints. She also had an impressive spread on the ground floor of the most beautiful blue skyscraper I had ever seen. I took up residence for a while, visiting it for art shows where we would offer amazing buffets and wine in a fabulous place to hang my art, inviting many notables in the environment and the Vancouver scene. It was there I sold some of my most expensive pieces for prices as high as $18,000 and $25,000 in the 1990s. We also had John Denver (my idol and composer of 'You Fill Up My senses' the song we married to) stay there in the apartment hotel rooms upstairs. That was for a big event where Rick and Donna, our friends from Gibsons and heads of B.C. Spaces, arranged to present my original to Premiere Harcourt. It was of a Glacier Bear in the Tatshenshini, an important area bordering B.C. and Alaska, which was protected by B.C.'s premier Harcourt at the time. John was to sing at the event, along with Loreena McKennitt. The event was held at the Orpheum, and I was to present the original on stage. That was probably the peak of my art career.

Life was busy and fulfilling and fun with galleries all over buying my limited edition prints, art shows with notables like Robert Bateman and Bev Doolittle, articles in magazines, and a taste of the high life.

Rein in front of his Portrait

In Motion

Commanding Presence almost understates the powerful image of the Grizzly that dominates the work that recently won **Liz Mitten-Ryan** the B.C. Wildlife Federation "Artist of the Year" award. Mitten-Ryan is a native of Vancouver who trained in London and at UBC. Her work is strongly spiritual. "I believe that nature is speaking to us. It's in nature that I find life's answers," she says. "I try to recreate moments in nature that awaken in me a sense of peace and joy, themes that help people recapture moments of love." This year Mitten-Ryan took control of the business side of being an artist by starting Silent Whispers Publishing. But even in business she is true to her spiritual nature — the company's name comes from the Navajo legend of the spirit quest, and refers to the silent whispers, the lessons, found everywhere in the solitude and wisdom of nature. She has shown her work across Canada, in the U.S. and in Europe, although you can find her work at Bernadettes, Farquhar Place and Barr Gallery among others in the Vancouver area.

Artist Liz Mitten-Ryan stands before a Commanding Presence.

Liz winning BC Wildlife Artist of the Year

Liz fundraising to get whales out of captivity

Liz, the Wildlife Artist

Liz presenting a Glacier Bear in the Tatsenshini to Premier Harcourt after the area was protected

Liz and Kevin make headlines when they move to Stump Lake

Chapter Seven:

FOLLOW THE YELLOW BRICK ROAD

In our farm life, we had gone through a series of farm animals so we could feed the family and write off the taxes on our acreage. One by one, the girls all became vegetarians as we couldn't eat our friends. Even Peter gave up eating pork for seven years after falling in love with the pigs from several attempts at riding them, and he's practically a carnivore. I eventually had the bright idea to breed horses that we would never eat but sell one baby a year to meet our tax requirement. It sounded wonderful as I could immerse myself again in my favorite animal.

It started with L.E. (Limited Edition), my first warmblood mare. I found an ad in a horse magazine, and the breeder was in Kamloops (about ten kilometers away from our future life). Being heavily into barter in those days, I asked if he would trade for art, maybe a portrait of his stallion. I took up my portfolio, fell in love with my horse, and, a few weeks later, had secured my first broodmare. She was only two years old at the time, so I had a year to train her before breeding. Remembering my Chako days, I proceeded bareback with a halter, and L.E. was a gem. She was, after all, a higher being sent to me to begin my next - not yet conceived of – adventure. When the time came, I picked a stallion to breed to, and again, my portrait offer was accepted.

Lynx was a stunning black Holsteiner that competed at the highest levels. At four years old, L.E. gave birth to Prima, who now at age 22 has been the love of my life. Diva, L.E.'s sister, became my second mare, and along with Mira, another sister, it was getting crowded on our smallish property in Gibsons. There really was nowhere to ride except along the power lines up the hill behind our house or along an increasingly busy road down below. I spent a lot of time leading or riding them to grass as they were denuding our meager pastures.

An interesting aspect of our property was it must have been some sort of portal for lost spirits as our children told stories of 'poncho man,' a native American that was often sighted. Cara had a couple who would walk toward her from the corner of her bedroom.

I had exchanged another painting with a psychic who agreed to visit one day a week for a year and coach me in her craft. There were many interesting stories around that, like the day she called in some cooperative spirits and taught me to find them and send them to the light. Kevin came across the deck from our office to cross the living room and have lunch one afternoon to be told: "stop, you're about to step on someone." That someone was a bit of a trickster like the imaginary friend in the film 'Drop Dead Fred,' and he was lying on the floor. Well, Kevin walked around him to the kitchen, and I found him and sent him to the light.

Also, there began a series of visitations involving me that happened several times a week for an entire year. I would be dreaming, and in the dream, someone was trying to get my attention, and I would wake up to find a spirit hovering over me. It was a bit disconcerting, so I asked God to reveal what it was about and to let me know if they were good spirits there to help me. That night I had a dream that I was a child in a swing, and the offertory hymn was playing "Praise God from whom all blessings flow. Praise God all creatures here below…" I woke up to an iridescent white light sort of in body form hovering above me, and I said, "come on in," and its spirit melded with mine.

I had dreams of funny animal beings like someone in a dog costume, and they would push me down on the couch and lick me all over. I would wake up, giggling. One dream was so wonderful as I was connecting with a big group of people I had known and loved from many lifetimes, and I woke filled with gratitude and with my arms in the air with many, many spirits all who I loved reaching out to touch me. One I call the tunnel dream as I was standing in front of a big tunnel. I was next to God, and he told me my life was over. I remember the finality and the feeling of trepidation, but I was with God, and that was what he wanted, so I said, "I love you, God," and went through the tunnel. The dreams went on and on, and I feel that like Elaine, my psychic teacher, they were getting me ready for another phase of my life.

For Kevin, he would listen to my stories, and I'm sure wonder, but he took it all in in his usual calm manner. One night though near the end of our time in Gibsons, he had an experience. He was paralyzed by a white light at his throat, and he felt something very, very big, was on its way. He just lay there and absorbed an influx of images and emotions, much like being given a microchip of information.

Around that time, I began going to another meditation group, a group of my women friends who met one evening a week to explore our burgeoning spirituality. We began sharing the book 'Your Hearts Desire' by Sonja Choquette. There was an exercise to visualize your perfect life, which I still share today in my Equinisity Retreats.

I really hadn't had time to think about it before, but I quickly discovered that I was close, but as they say in the 'Little Mermaid,' I wanted more.

I dreamed of freedom and rolling hills and lots of grass. I added those dreams to my 'hearts desires' perfect day and sent it out to the ethers. It happened so quickly that it caught me off guard, and if you think it took me by surprise, imagine how it affected Kevin.

As seems to happen in my life, my commitment to be in my truth and listen to my intuition invited a host of happenings to bring my dream to pass. First of all, my brother, in a rare conciliatory visit, put his arm around me and suggested we walk and talk. He said that we, including all our joint children, were really too many to share Roberts Creek, our family summer place, and either he should buy me out or I, him. It actually worked out that the problem was solved by my father leaving it to Steve (which we agreed on) and giving me my share in funds to place a hefty deposit on my 'Hearts Desire.' Steve then took me to the internet. He showed me two new developments in the Nicola Valley, where existing large ranches were being subdivided.

Strangely, Kevin and I were set to go on our yearly week-long camping trip sans kids, so we went along with the suggestion. You know the phrase 'this or something better,' when we drove up to our now property of 20 years, our magical sacred land, there was nothing in the world, as far as I was concerned, that was better.

I knew it, I recognised it from that place where we view our lives from higher planes. I asked for signs and got them. The heavens shot light beams across our land, eagles dropped feathers and lined up, nine in Gibsons to say goodbye. There was no choice except by Kevin, who looked a little dazed and confused but agreed knowing he was on a new journey, and he better well comply. One of the first cards he ever gave me was a Far Side, a great cartoon of a frog with his tongue attached to a jet plane… It was too late; he had caught a big fly. That was 12 years into our now 32-year marriage. We had survived six children, even Peter, and we were still madly in love. It was going to be tested.

Just writing this now, I'm beginning to look at myself from a third-person perspective. "That woman is truly crazy," they would say, and I can agree. I didn't think twice. We had business careers, offices, studios, projects, products in the form of thousands of limited prints… How to get rid of them? There was a gallery in Alaska that was my biggest client. I had the largest room in it and went to yearly shows there. I would send all of my prints to them. That was easy. It actually resulted in my never seeing them or any proceeds from their sales again. Selling the house, which we had to do to complete on our land; that would happen. It was all under control (just not mine really). As we were driving home from our newly acquired project, I blurted out, "I think we will sell the house in June and be out by September." I have no idea where that came from; well, in retrospect, I do. We returned home to a couple who "just had to have the house." It was June, and the closing date they chose was September 18, my birthday! Signed sealed and delivered, we were on our way.

Interestingly they were a childless couple who just had to have our huge family home with teen house, office, barn, help house, etc. etc. Kevin had to do some shuffling and even return for several months after our move to tie up loose ends. Our youngest Cara had to live in the teen house for a few months until she graduated from high school early.

Chapter Eight

UP, UP, AND AWAY

We moved into a wall tent, a very rustic early version with a mattress on the floor, a leaky wood stove, nails for hanging things on, no water, no refrigerator, no bathroom or shower. It was wonderful! Also a bit of a change from our previously busy lives with friends, family events, community, to the middle of nowhere and absolutely nothing. Trees, hills, lakes, vistas, peace, animals (a cat, three dogs, and seven horses). There was nothing to do but walk the land, contemplate, meditate, and experience such gratitude. It was huge, and the future spread before me, a blank journal waiting to be written on, a spiritual journey newly embarked upon!

As the initial weeks of Indian summer and the thrill of freedom passed, there were a few details to dissolve. There was supervising the construction of our soon to be barn and home, fences, roads, wells, infrastructure, as we were to be off-grid, solar, and very rustic in our initial stages. We needed to get in for winter, and we finally accomplished that on December 2, which was a bit over the tent's due date. Our new quarters were a rudimentary shadow of our life of comfort and convenience in Gibsons. Its main appeal was warmth at that stage! I flew off for one last art show in Alaska and returned to a long, slow winter to hibernate and digest my new surroundings. I remember thinking everything was slow, like in suspended animation. I journaled, meditated, and kept the fires burning but plaintively asked for guidance on what was next. One grey day, I viewed a glimmer of color in the distance, which, as I focused on it grew bigger and brighter until it eventually landed right in my paddock. It revealed itself as a bunch of helium balloons that had run out of steam and, on inspection, had a sign that said, "You have won the shooting star award." It was from Hollywood Video Store in Hope B.C. That really said it all, so I just settled down to wait for inspiration.

When I had gestated, I put out feelers for possibly a new art career and was sent in the direction of Trudi Backman, my now friend of 20 years and final agent for my deflated art career. Painting the odd scene for galleries wasn't going to support either of us, so Trudi wandered towards interior design, and I, to my next creation. To this day, we habitually meet one afternoon a week for a recap and a glass of wine, usually accompanied by a lot of laughter. Interestingly, as this is book six for me, between Kevin and Trudi, they have announced the conception of most of them. For this book, we were reminiscing next to a wonderful painting in Trudi and Lyle's living room, which became the inspiration for the title, and I reproduced as my version to become the cover for this book. I was regaling them with silly stories from my past, after which Trudi announced it was time to write this memoir.

Life is an adventure full of fun and excitement when we discover our destiny and follow the prompts. We've really have no way of knowing what sort of influence our lives will have on others, so it is best to just listen and trust that there is an intelligence that does. So having introduced this as my sixth book, you now know that writing became one of the directions my life took. I surrounded myself in my growing family herd of free horses, learning everything I could from them, and the land, this amazing and natural reservoir of nature's wisdom. I learned about natural horsemanship, and the language of the herd, eventually rebranding it 'Natural Horsefriendship.' I rode the surrounding miles of hills bareback and bridle-less, walked the land, and began to channel the wisdom of both.

We were always growing and evolving as more and more people were being guided to join in this experience. Some came in as shamans and geomancers to teach more about the energies of the earth and help guide the early retreats. Some came eager to absorb the offerings after reading the books. We began to grow the retreat center with guesthouses and guest wall tents. We built a spirit lodge for ceremony and a healing temple for healing and sharing, and more people came. We had people show up to work here and built a help house, and the work increased with additional horses and guests.

The help has been an ongoing conundrum, as the people reading the books and watching the films (which came next) generally knew what to expect, but the help really entered a strange land that didn't really relate to any previous employment. They had to hibernate and do a lot of shoveling in the winter and emerge in the retreat season (May to October) in hospitality mode. We have an ongoing parade, some are rejected by the powerful energies, some implode, some remain friends but move on, but most I believe come for some sort of awakening or readjustment and move on without totally appreciating what just happened.

One constant in my barn help over the years was my four grandkids. Daughter Alex who had disappeared to Nova Scotia for years as her husband Colin had gone to university there, found a house in Kamloops when the four grandchildren were age three, five, seven, and nine and over the years they've been the best ranch help coming on weekends to get spoiled by Kevin and I and help us through all the continuing sagas with the ever-changing help. They've trained the help, cleaned up after them, and could all basically run the ranch better and faster than any help so far. It's been wonderful having them so nearby.

We have many happy memories of big Christmases and gingerbread houses (architect-designed to blow up during mock sieges) sledding, skiing, ice skating, along with all the horse activities. They have (at least Kaia and Thaelo) succeeded in doing-in two Kubota's while learning to drive, and run the retreats (Imy and Rhea) when help disappeared. The sad thing is that they all grow up and go off to post-secondary education too soon.

People often ask me how my grandkids got their names. Their mother wanted to give her children unique names with a powerful spiritual meaning. Kaia means 'Gift of the Earth, Flows from the Earth' Rhea, basically the same as Kaia. Thaelo was even more imaginative. Alex called me to ask some names for the color green. I offered Veridian, Emerald, and Pthaelo, which she chose, but thank heavens shortened it to Thaelo. The rest of his name is Ari, Ralph, and together they mean 'Unearthly Green Lion sent by God to Council Wolves.' Imogen is basically 'Imagine,' and we call her Imy for short. (I, My, I, Me) that's her. A funny story of their names happened when the children were still very young, and a washing repairman was called. The four of them gathered to watch him work. After some time, they were curious as to their new friend's name. When asked, he said, 'Bob' and they answered in unison… "That's a weird name."

I just had a recent conversation with Thaelo, now 17, whose hero is Elon Musk. He was sharing that Elon had a new inspiration called 'Starlink,' which is a global satellite network that would be high orbit and higher speed and would possibly trump 5G. We started to talk about some of Elon's ideas like the electric car, which would steep its drivers in higher EMF's. I said that possibly the Starlink would do the same I shared my understanding that as consciousness reached higher levels, we would not need to be subject to the limitations of time, space, and dimension, the problem of EMF's would be solved. He was also sharing that France and Germany had just invested in a super battery plant. I volunteered that space travellers from far away galaxies most likely had solved these problems and were operating from higher levels of consciousness.

Really, I always look back at us from a higher perspective and see we are really still in the dark ages where we are viewing those shadows on the cave wall illuminated by the campfire.

Our retreatants are a totally different story. They're genuinely called and led to the perfect time and experience, very often with kindred spirits. They seem to fulfill the purpose of all of us, the land, the herd, myself, to guide them on a spiritual journey to discover their calling. Once the connection is made, it remains in place. They have a new higher level of resource that is ongoing in their lives once they have made it. Their hearts are opened, and they become the flow. It is so profound to see time and again; the horses show up in different ways using sacred geometry, rhythm, flow, and their ever sensitive tuning devices to read and reconfigure the energies. The people show up for each other sharing common themes like wanting to open their hearts or take a new step or learn to trust or heal from trauma. It is a privilege to witness and be a part of the program. There have been many magical occurrences in our twelve years of retreats, along with many things we still don't understand. I feel we are always headed to a new level of understanding.

View from the front pasture

Nono, Liz, Denny, Leslie, Dana and Belle at Stump Lake for Liz's 50th

Kaia and Chiara, first-born grandkids

Kaia and friends

Gabriel and Prima

Kaia, Rhea and Imy on Tesoro

Kaia and Gabriel on Merlin

Imy and Prospera

Gabriel, Imy and Thaelo with Merlin

Paschar and friend

Epona and Paschar visiting mom

Epona and Paschar visiting mom

Liz riding Prima

Kaia, Rhea and Merlin

Thaelo and Tesoro

Kaia's first ride on Paschar

Liz receiving two IPPY awards at the L.A. Book Expo

Chapter Nine

APRIL BRINGS THE SPRING... AND NEW CREATION

The first years on the ranch were spent living simply and growing slowly. The horses were all-consuming with breeding, birthing new foals, training, and just general immersion in the horse culture. My friend April who I had met through Choices, which I talked about in 'One With The Herd' (her husband Clive was my T.A.), was very interested in exploring the horse world as well, so she would often visit and stay for a week. We studied Parelli through Jonathan Field (whose father bought the ranch next door) and, together with our other friends Shirley and Jane, moved up through the levels becoming better and better at our horse skills. I still maintained my bareback and halter riding and got very good at galloping through the hills. It was so liberating and fun going exploring as you can ride for hours without finding a fence. It was also a good place to train the young horses, and together with my friends, we would explore the incredible landscape of hills, forests, and lakes cloaked in endless views.

April was also my birthing buddy and saved Serene's life by giving her mouth to mouth when she was pulled out by the vet after a difficult birth and was not breathing. We had many sleepovers in the birthing stall and special moments welcoming new life into the world.

We also shared the vegetable garden, which grew with each passing year. April was the garlic queen and has continued to perfect our large and flavorful garlic. When Mira's first baby Luxy was born, April was in love. Luxy was a joy with her patient and gentle manner and a beauty with her four white socks and a star. She and April had a love affair until her death at 13 years old from severe colic. Miro is now the only one left from that family, and he has the same lovely temperament and good looks.

We began inching away from Parelli, which was based on pressure and release, into the horses' idea of a more inviting exchange, which they called 'Natural Horsefriendship' or 'Invitation and Reward'.

L. E. was the teacher, being the beloved and wise lead mare and grandmother spirit of the herd. She taught focus and intent, which was clearly demonstrated by her behavior in the herd. She would walk through, and the waters would part. If she wanted something, she simply went straight toward it with her daughter Crystal in tow, taking full advantage of her mother's position. We began to teach people to be lead mares like L.E. and used more and more rewards and liberty play. The carrot sticks were replaced by wands, and the levels of pressure became lighter and lighter.

I had been engulfed by my new life for several seasons now, just absorbing and becoming renewed by the gifts of nature and my growing Herd. I would do the odd painting, mostly scenes of horses or cows spread across the landscape. It was no longer a primary focus, and I could sense something larger, gestating from my immersion and connection to this new life.

One morning over breakfast, Kevin looked at me and said: "When are you going to write your book?" "What are you talking about?" I answered, having not the slightest idea for a book in mind. Apparently, it was the elephant in the room, too large for me to see. The next day, I sat down with a pad of paper and began to write. I wrote and wrote as long as the channel was open, and then I was done. For two and a half months, I continued until it appeared. "One with the Herd" was finished.

My son Peter who is great on the computer, helped with photographs, graphics, etc. and got the book to manuscript form. I remember so many funny incidents from the process.

One time, I was laughing so hard, I almost died. He was cleaning up a photo of L.E. with Kaia in the snow with me standing a few feet behind them. There were little piles of manure on the ground, and at my request, he just photoshopped the offending objects out of the background. While I was in the kitchen preparing lunch, he secretly decided to move a pile of poo to the top of my head. Then he added a speech bubble from L.E.'s mouth saying something like, "Wow, that was a big bird." I was in the middle of swallowing a handful of vitamins when he showed me, and in my immediate laughter, inhaled several. I began jumping and gesturing that I was choking to death, and he just thought I was reacting to the joke. I finally threw myself over the back of a nearby armchair in an attempt to give myself the Heimlich maneuver; the pills rolled out, and I'm still here today.

I didn't know the first thing about book publishing, so I got books on it and sent off manuscripts until I found a publisher. Strangely, other than copy editing, they didn't change a thing, just helped with the layout and design process. It was a very expensive book, in the end, a hardcover with a satin ribbon for a bookmark, and it was perfect. The art and the pictures, the type, the paper quality, were all the best, and it was sent back to China, where it was printed several times during that process until they were.

We won five independent publishing awards for it, and I still have boxes of book award stickers that were never applied. I learned a lot, and then on to book two, three, and four. I wrote four books in two and a half years. Then I went to readings and book shows and finally the Los Angeles Book Awards to be presented the nine awards I had won overall (I stopped submitting towards the end since I wasn't going to use the stickers anyway). The L.A. book awards was an adventure in itself. I stayed in a hotel by myself, actually found the venue and help just materialized when I needed it. I arrived at the door with several boxes of books and some kind behind the scenes helpers provided a vehicle and helped drive me to and from my booth and appointments.

I walked outside the building on award evening to find a three-block lineup of people and cabs as we had to go to some high-rises for the gathering. I was not sure exactly where but help again materialized in the form of a woman who took me by the arm and accosted the first cab (after pushing the two of us to the front line). She then told the cab, "do you know who she is?" and where to go. I looked across the back seat at her, not even asking who exactly she was, and we wound up at the top of the very posh high-rise and again at the front of a line of waiting people. I think she was just a gutsy party crasher! There were people there in evening wear with their families and friends, celebrities, and me in my jeans, a tank top, and jacket. I spent some time talking to some musicians about my book and horses, and they brought me wine and kept me company until the awards were called. My gate crashing friend later asked, "do you know who that was?" of course, I didn't, and the name Guns and Roses and Axel Rose didn't mean anything at the time either. Now, if it had been Leonard Cohen, I might have left Kevin for him. They were nice either way, and we ended up on the podium together to receive book awards. There were Indies and Ippys and Nautilus, and I was the only one to receive two "Ippies," one each for different categories of the same award. I had hugs and too heavy metal coin-like things on ribbons placed around my neck, and it was over. There were more awards the next day back at the book award building and really the Nautilus was the most prestigious with the company of some great writers like Eckhart Tolle, Deepak Chopra, and the Dalai Lama, who had won alongside 'Prima' my horse who had been recognized as the author of 'The Truth According to Horses.'

My cousin, Sandy, who had been kicked out of nursing school alongside me, had been living in L.A. since. She had been successful at husband-hunting and luckier than me as she married her first love a successful advertising agent in L.A. We had a fun night and overnight at her house laughing at our exploits and many fond memories of growing up. The next day I was off to meet Carolyn Resnick in Escondido outside of San Diego. And I had no idea that there weren't buses to get there. Then there was a hair-raising ride with me driving in a rented car with my first GPS on a six-lane highway, but amazingly it got me to Carolyn's doorstep in one piece. She and I had met previously and were on the same page with herd language and our complete absorption with horses. Carolyn had worked for years with difficult stallions coming up with ingenious ways to get them to see her as lead mare. One story she told me that I will never forget. She was having problems with a particular stallion that didn't respond to her usual methods, so she had to think outside the box. When she was given a gray elephant squirt gun from McDonald's, the idea came to her that this is how she would show him she was the boss. After weeks of trying every other method, she walked up with her little squirt gun and gave him a little squirt between the eyes, and she had his respect from then on.

I forgot to mention we had actually met because of my art. Carolyn had fallen in love with one of my horse paintings, and that allowed me to begin the friendship when I first delivered it to her. We also both had Summit Drive addresses, so we called ourselves the Summit Sisters. All of my books were put on Amazon eventually, as that seemed the easiest delivery vehicle worldwide. Gradually people who had read them began asking to visit, which eventually turned into our retreats.

Chapter Ten

ENERGETICS OF THE LAND AND MODERN CONVENIENCES

Our house had remained rudimentary since its inception now 12 years earlier. It had a tiny propane stove and fridge, which were years behind the times, and as the retreats started to happen, were inadequate to handle the growing numbers. The Guest House had been built and later the Hill House for help. We even had the tent village with its shower house and Spirit Lodge, but Kevin and I were still camping in our Barn/House. I had invited Christian Kyriacou, the House Whisperer who I had connected with in London through his brother, my old art school mate. Christian was fascinating with his knowledge of Feng Shui and sacred geometry, and he was often invited to ghost bust as well. On our new land, we had run into a few nature ghosts, and I was interested in his interpretation. There were so many layers being discovered from the geomancy of the rocks, trees, and energies, to the historic native elements. We had begun to map out all of the most intriguing places and continued to find more. Three hundred twenty acres of forests, hills, rocks, lakes, and meadows is a lot to explore, and I began to develop a friendship and connection with my land. The Merlin tree in itself holds many mysteries. It is a prophet and a clear connection to higher consciousness. I would lean against it, and it would give me a vision or a sentence relating to what was ahead.

One year, it showed me filmmakers coming from three directions, and the next morning I had an e-mail from a filmmaker, "can I come and film?" That summer, another woman from the biggest TV station in Brazil arrived, then Lisa Lightbourn came and filmed our first documentary titled after the first book 'One With The Herd,' and we won best documentary for it. Another year it showed me the Disneyland Castle with Tinker Bell flying around. I was really stumped, but that fall, Kevin surprised us with a trip to Disneyland for the grandkids and us. It wasn't until I was standing in the same picture that I remembered the vision.

One of the best Merlin predictions was when the cranes were newly arriving in the spring. They fly in, in the hundreds, and are just black specks high in the sky, but they have a characteristic call that is always a welcome sign of spring. I was sitting by the fish pond, looking toward the Merlin tree on the hill behind, and the cranes flew over. They dropped in the sky and surrounded the Merlin tree swarming there for several minutes in the hundreds. Then two hawks flew towards me from the west. I was not sure exactly what it meant, but I knew it was good!

Anyway, back to Christian. We had invited him to come and help decipher the energies of the land and also to consult on our house, which we were finally going to build. He brought over some dowsers who use metal dowsing rods to interpret the earth energy, and they began with the native energies. Apparently, this land was a border between the Shuswap nation to the north and the Okanagan nation to the south. They found marker cairns, fire pits, historic tent villages, and a beheading rock where I had been beheaded in a previous life. The story was that I had a previous life here as a Shuswap, married daughter of the chief. The love of my life had left for some time, and an Okanagan had violated me. On my husband's return, I was beheaded as I could no longer be with him. I could see the vision, but I went to the Light, forgiving him, and still loving him. Strangely it made sense why I had been called so strongly to this land. Whenever I would be returning from the Spirit Lodge and get in my truck at the bottom of the pathway, my husband would climb in the back seat and ride over with me. When Christian was staying at the guest house, he would do the same with him whenever he passed that spot. He wanted to be forgiven, and we did and sent many natives to the light. The activities around the tent village settled down, although people occasionally see him or the chief (my father in that life) today. After his clearing the land help, we settled down to plan the house.

Kevin (an architect), Christian (also an architect), and I meditated on the form the house would take, and it took on sacred geometry around our core. We still have a place in our den that we sit every evening, and the house was designed in a Nautilus around. The Nautilus became the living and dining room with an octagonal bedroom wing all surrounded with tall windows and sliding doors out to all the porches that face south, west, and east. Directly below and beside is the barn and horse paddock so we can walk down some stairs and be right with them. The house has wonderful energy that flows through its open plan, and every view is a picture of course, of my favorite beings, my horse, and land family.

After enjoying our rustic existence happily, and even dragging my heels around building our dream home, I was all over it once we were on the path. When we had left Gibsons, we left behind even the most basic comforts like a dishwasher, built-in vacuum, convenient heating, laundry and deep soaker tub, telephone (for quite a while), a decent size fridge, stove, etc. etc. For our 'dream home,' we needed to upgrade our solar, wind, and all the paraphernalia associated with off-grid living. For that, I had to sacrifice two-thirds of my tack room, but I think it was worth it. Our new home is so much more comfortable except on the odd occasion in very adverse weather like -22 degrees Celsius when our infrastructure fails. This could be a broken fan on our in-floor wood heating, or a small oversight in Kevin's programming the generator like not putting it back on automatic. I'm just remembering our latest incident when our heat dropped 10° very quickly. Our heat pipes froze along with several sensitive items in our pantry (like the last of my coveted organic garden potatoes). After such an incident, it can take days to regain equilibrium, and it involves more hardship that I would welcome at this stage in my life.

Friends say, "why don't you just put in hydro," but at this point, it is really against our religion having worshiped in the off-grid denomination for over 20 years (and invested countless resources, time, and money in its perfection).

I do hope we work out all the kinks before our retirement, which is looming now that we're both approaching our seventies. The inspiration is to eventually build our retirement home and sell the property and retreat business to a younger couple who inherit the philosophy, love, and commitment to this consciousness and connection to the land, horses, and retreats. We will stay on as advisers until our demise. I hope they don't mind dragging my shell up into the hills to feed the animals! I'm organic, and I've spent my life feeding animals, so why stop then.

Of all my children I have shared that maverick view with, only daughter Alex was on board. But then, she was grateful to eat Tesoro when he died as I could hardly even feed his beautiful organic 12-year-old beef to my dogs. I loved that steer and will miss him forever! Bless you, Tesoro. Tesoro, for any of you who don't know, is the wonderful white-faced black Angus steer in our award-winning film 'Herd' on YouTube. He was a higher being and one of our greatest teachers here at the retreats.

We have since enlarged the Guesthouse, adding a two-bedroom east wing and more wall tents to our village. Last year our Healing Temple was created. It has a pyramid roof surrounded by sliding doors like a gazebo. The pyramid has an interesting story. I began to be fascinated with orgonite a few years before, which uses casting resin and layers of crystals and metals to make a healing, energy clearing, beautiful piece of art. I had experimented and researched my crystals and created several small masterpieces, some of which I kept, and some I gifted.

I had the idea to make an 8-foot pyramid roof for the Healing Temple with colored crystals, which would use both color therapy and the healing properties of the crystals. It was going to require gallons of resin to affix the crystals and stained glass and be a mammoth undertaking. Still, my creator self was already in 'I have to do it' mode. I ordered all my supplies, and they arrived box by box. Garnet, Peridot, Lapiz Lazuli, Quartz, Rose Quartz, Turquoise, and multi-colors of stained glass. It was time to begin as the resin needs a warm temperature to cure, and we were getting closer to May when the retreats would start. I was going to heat the east wing and execute my artwork there.

That was until one morning I suddenly looked at Kevin and said, "how will we get an 8-foot pyramid through the door?" There was silence. We had a glitch. I racked my brain, thinking, "where would I be able to do it?" Possibly a warehouse or shop of some kind, but I needed heat, at least 21° C of heat.

Suddenly it occurred to me, my friends Trudi and Lyle had a heated two-car garage! Trudi (and Lyle after some pleading from Trudi) said yes, and I was off to the races. However, it was a mammoth project involving four men to come and turn it each time I needed to work on another side. It also involved Lyle and Trudi parking on the street for some time, and Lyle not being pleased with the resin drips on his pristine cement floor. It did sorely test our friendship (not from my side) and involved the biggest most expensive bottle of scotch as a thank you to Lyle, and will probably never be allowed an encore.

Chapter Eleven

RETREATS

I briefly mentioned earlier my 'Choices' experience when Kevin and I were living in Gibsons. We enrolled in a 4-day encounter group designed to break down the barriers to new personal growth. It was there I met Clive and April. I sorely tested Clive, my Teaching Assistant (T.A.), whose job it was to corral me into following the program. It was difficult, and there came a time when I had burnt out Clive, his higher up, the next higher up until I finally reached the minister who headed up the program. His name was Forrest, and he truly was a man of God. I know we could see each other amongst the 100 or so participants.

One day on my lunch break, I had bent over in my hotel room, and my given name tag of 'Master Manipulator' had fallen out of its plastic sleeve and was blank side up on the rug. Inspiration had me write GOD on the backside and rebrand myself. Well, when I returned to the room, it was viewed as blasphemy, and I was crucified! When Forrest was called, I was in the middle of a circle surrounded by several angry people. He asked me to defend myself, and I shared my knowing that we were all, in fact, GOD.

He agreed, and later, he baptised Kevin and re-baptised daughter Francesca and me in the Hotel pool. He and another seer from the group told us that Kevin and I were anointed and would one day be surrounded by many people. I was also told that I was preparing my galleon for launch and stocking it with all the provisions I would need for the journey. (This was shortly before we left Gibsons for Gateway) At the time, Kevin and I looked at each other with a mix of interest, woven together with knowing. There were a few adventures still forming their roots in our soon to be new soil.

Our first retreat came together, organically after a shaman arrived at the property. I can't remember the details, but I think she was introduced by my friend Loesje who I had met when I hosted a Body Talk for animals and humans retreat in our early years long before there was any more than our Barn House. For that, we rented a log house at Stump Lake Ranch below and organized practitioners using the horses and the land here for some of the teachings. Loesje was one of the participants who signed up, and right away, we became friends. She has since gone on to historically be the only instructor for Body Talk for animals in B.C., also developing her own system Linking Awareness, which she now teaches all over the world. Anyway, our courses were really fun and have influenced some of the healing techniques I continue to use particularly Body Talk Fast Aid, and its amazing five treatments that can be taught in a weekend. The course ended with the creation of a sacred circle that we created on the land and is still there today. The horses watched us work on it from a distant hill. When we had finished our ceremony, they promptly ran down and through our circle, tipping rocks, eating cornmeal, and claiming their sovereignty.

Billa was the shaman at our first retreat, and she was a red-haired, freckled (definitely not native-looking) woman, maybe in her late thirties. She walked the land in awe, finding and explaining some of the amazing energetic areas. She explained how to see vortices in the surrounding growth, which becomes a spiral, and together we mapped out the chakras on the land. The land has a full set of chakras like on the human body, starting at the Earth Star (the chakra below our feet, the last part that is us, before we leave our bodies and connect to the core of mother earth) to the Trans

Personal or Individuation Point which is the last chakra above our heads where we enter and leave our bodies when we're born, die, or astral travel. The rest from the root to the crown, are arranged almost like on the body from the root at one end of the property to the crown at the other. It really opened my eyes to the energies of the earth from the rocks and trees to the overall pattern of energy represented by different landforms.

We have dragon energy at Gateway, which offers metamorphosis and manifestation. Of course, that is transformative in the retreats. Billa and I planned the first retreat together, and building began on the Spirit Lodge, which is a native pit house used for our ceremonies. It is the hub of the land energies and past native histories. It is mostly underground with a doorway into the hill and a skylight on its roof. The smell inside is wonderful, and the feeling one of welcome and connection.

We then built the wall tents, which are a wooden frame with wood floor and walls to four feet high at the sides, and canvas rooves. They are very comfortable with beds, and more like cabins than tents. Then there was the shower house and outhouses to finish off. Our village was tucked at the end of the trail through the old fir trees and next to a lake and a meadow where the horses often sleep. It really feels like going back in time. The trees around create circles and inviting little clearings with some double or even triple trunks reflecting the powerful energies there. Owls and deer visit, hawks, and coyotes with the occasional native spirit just dropping in from another time (as I mentioned previously in my past life story).

Our first retreat was filmed and named 'Equinisity,' which was a joint venture between Frank, the filmmaker, and my work of editing (or rather not editing enough). It wound up at a 3 hour running time including the behind the scenes footage. The people called to that retreat were an Australian who had a large retreat center there, Jane, Kerry from New Zealand, a Bowen therapist, a psychotherapist from New York who discovered that horses were the best psychotherapists, Connie from Washington who has gone on to become a close friend, a photographer, our shaman Billa and Dana an EFT practitioner, Carol, Billa's mother in law, and energy healer, and Frank the filmmaker. I think we did a pretty good job with the story and showcasing all the special places on our land journey.

The horses healing was in its infancy as we were just discovering the possibilities. We had built wooden healing tables which people laid on, and Billa or Carol or sometimes another would hold space at the head of the table, and the horses would come over and help do the energy work. Over time it became restrictive with the people actually discouraging the horses is by getting in their way. My two facilitators also developed disagreements over different aspects and bottom line; it turned out the horses could do better without them.

We evolved from the healing tables to letting the horses choose the space and moving stools nearby, to our plan today, which is an open barn with eight seats built around the center posts. This works best as it allows the horses to move where they will using sacred geometry and patterns, opening hearts, and offering their healing energies alone or in groups.

We began learning more about the land as people began to have their own experiences. I remember Susan, who discovered what she called 'the Crystal Children,' who she would see come out of deep holes in the earth. The land has rivers of underground crystals, fluorite, chalcopyrite, and quartz, which are all healing crystals, so it made sense. People would all see different aspects of the land and tended to come in groups of kindred spirits. We had one retreat that was comprised of geomancers, dowsers, and energy sensitives who all claimed to have been here in other lives. One of the people apparently was my native love who beheaded me in that life, and she couldn't go anywhere near the beheading rock without feeling sick. Another saw natives all around the tent village. We have people who see unicorns and fairies. Izzy, one of my previous barn help and now unofficially adopted daughter, would always see fairies, some large about 4 feet tall, some very tiny ones used to pull the sleeves of her top guiding her in the throat chakra area.

I believe everything in the human imagination is available in its dimension because time and dimension are just aspects of our understanding of reality, and there is so much beyond 3D. We have had very normal people like our English couple, now friends, who were walking to the toilets at night. He saw a native chief in full headdress (my father from my past life), and his wife observed him while not seeing anything at all. We had the poltergeist episode one summer. The oldest tent in our tent village was getting a little tired from years in the sun and was very nearly ready for replacement. I say this as we tried to simply put the stories down to that. It started with a woman staying there telling stories of knockings and of crystals which she had arranged on the window ledge being knocked off. Then another in a nearby tent complained of similar signs and then another on the other side. It was primarily centered around the old tent. After the retreat, Kevin and I went over to spend a little visit in the tent and assess the situation. I asked him if it was possible; it was the wind on the tatters. There was no wind when we were there. Then I asked if it could be something in the wood structure needing repair, and that was a definite no. So we decided it was time to order a new tent and Kevin left to get a measuring tape.

I lay down on one bed and communed. Bang, bang, bang, bang! A loud knocking came from near my bed. Huh? Again there was no wind. A shaking on the other side. All right, that was too much. We ordered a New tent anyway, but I knew the problem was bigger than that. Richard, the geomancer from Montana, was due to visit and had originally appeared in our lives when he called me from Montana, saying that he had been drawn to our land through Google Earth and that it had some amazing energies. I'll briefly tell that story for context.

At his first visit, I had only spoken to him on the phone, and my friend Trudi who was visiting at the time, said, "Who is this guy? He's coming for a visit, and you know nothing about him?" We went to the Internet and looked up his website, and there he was, the perfect picture of an elf (a big elf). We both looked at each other in anticipation of an interesting adventure. Well, Richard arrived and admittedly agreed we were right. He has some amazing powers and has continued to send me some very interesting energetics around the land. Like his rendering of an area that was emitting an energetic that looked much like multicolored shafts on an umbrella radiating out to touch various areas on the property and beyond. I immediately identified the location to be that of the Merlin tree). So Richard was invited to have a look at the tent. There was Richard and the three of us. We three sat on the beds, Richard entered and immediately said in his most authoritative voice like talking to bad children "This is not Okay! I want you to leave, go to the light and stop ruining these people's retreats!" he was, in fact, talking to some mischievous native children who were enjoying a little campsite raid (I remember my cousins and I doing the same to the neighbors at Roberts Creek when we were children).

That was the end of that, and the backup came when Loesje, my Linking Awareness friend, arrived with a group of doctors and vets from Indonesia for the second visit that summer. She told me the first visit one of her group had spoken of a native child that had followed her home and had been very difficult to get rid of. Never a dull moment! The native sightings have slowed down to an occasional visitation, usually when someone is doing ceremony in the Spirit Lodge as the chief will grace a particularly powerful invitation, but so will the owls, the deer, and horses. Talking about that, we have had great horned and great gray owls nest around the Spirit Lodge, their fluffy white babies sitting on a high branch and talking to us. They will sit on a fence nearby or even on the back of my truck. Last summer, we had a great gray owl regularly come to the fence beside our new Healing Temple to grace it with his presence. He would let everyone near to take close-ups. After the lightning episode and the end of the three witches visit, he brought his mate, and the two sat at either end of our Gateway 2 sign to let us know all was well, and they were on duty.

Chapter Twelve

LIFE LESSONS, KEVIN STORIES AND LATER YEARS

One of life's lessons that I have come to a good understanding of is money as flow. A product of my years of immersion in animals and nature is my knowing that the lilies of the field, the birds and bees, and all of creation even humans are all provided for. There is always enough. No amount of hoarding paper or shiny things increases that providence. Humans always get in their own way with their belief in limitation and hard work. Uncle Ben (Franklin) was a bit misguided in his quote, "A Penny Saved is a Penny Earned," and that kind of thinking has kept people enslaved for all time. As Kahlil Gibran so beautifully says, "Work is love made visible."

"And when you work with love, you bind yourself to yourself, and to one another, and to God.

And what is it to work with love?

It is to weave the cloth with threads drawn from your heart, even as if your beloved were to wear that cloth.

It is to build a house with affection, even as if your beloved were to dwell in that house.

It is to sow seeds with tenderness and reap the harvest with joy, even as if your beloved were to eat the fruit. It is to charge all things you fashion with a breath of your own spirit, and to know that all the blessed dead are standing about you and watching."

Once you understand that our limits are all the product of our beliefs, you become free to love and dream and listen to your promptings. Knowing that is how creatures that follow their hearts (and intuitive guidance that flows through the heart) are far beyond us in their wisdom. The horses have shared about this in their five books. If we could get past measuring our balances in all those connotations, then we would know our "cup runneth over" and possibly cut out the middleman of the exact exchange represented by a complicated and crippling system. I feel money as it is now being accumulated in the billions, and trillions has gone beyond a coherent system with any real meaning at all.

I always avoided math and found it a brain burn. Why bother? Things just are, without puzzling through the equation behind them. Like in quantum physics, everything can be proven in the language of mathematical formula, but that is for those who prefer to exercise the circuits in their brains. I prefer knowing to proving. Science and medicine are in their infancy, and there are so many unknowns that are limited by our intelligence quotients, again one of those many measures of our limited understanding of our full potential. "Knock, and the door will be opened," "As you give, you will receive. " "Believe, and you will move mountains" and "the Lord is my Shepherd, I shall not want."

All true pearls of wisdom worded for children to understand. We're children still in our school years and just beginning to evolve towards our full potential as Creator Gods, one with the entire larger body of God. I was invited a few years ago to participate in a book that was being compiled by Nicolae Tanase called 'Love-The Ultimate Answer to the Meaning of Life.' I was asked to send in my understanding, but as my understanding is nowhere near complete, I asked my Herd that always connects me with 'One Consciousness.' Here's what was included in the book:

Liz Mitten Ryan and the Herd, from "LOVE, The Ultimate Answer To the Meaning of Life" by Nicolae Tanase

LOVE loves and by its very nature is so vast, incomprehensible, and dynamic that as all of nature, we must simply surrender to its beauty, majesty, and wisdom to live as one with its intuitively directed flow. We are then one with the conscious energetic field of God or the ALL upon which we are all fleeting expressions existing in a state for a moment in time, and LIFE is the continuous expression of us all. In 'Life Unbridled, What Animals Teach us about Spiritual Truth' the animals share that L.I.F.E. is 'Love In Finite Expression,' forever.

Other synonyms for 'meaning' are definition, explanation, interpretation, understanding, and they reflect the human need to intellectualize and understand. When I asked the question of why I was given the 'Creation Story' written through the animals' perspective and on receiving it, I remembered its truth. We are all the meaning of Life.

The Creation Story:

"In the perfect realms of spirit, a group of creator Gods, with blessings from the Oneness or the ALL imagined and conceived a perfect planet E.A.R.T.H. (Expressing And Recreating The Harmony) Also the same letters in H.E.A.R.T. (Holographic Expression and Re-creation Tuning Device) which would contain a living evolving

library expressing the great beauty and diversity of Love In Finite Expression (L.I.F.E.) all in harmonic alignment, all interdependent, all one in the body of the ALL. (The Garden of Eden story.) The creator Gods descended from heaven to play in their creation and returned to spirit at will. Over time, the density and laws of time and space drew them deeper into their creation until some were too dense to return to spirit and chose to be part of a great experiment to see if the seeds of the Oneness planted in the good earth would ultimately grow to become the fruit and flowers of that Oneness.

The animals minerals and plants maintained a clear connection to the Oneness while the 'humans' as they became when planted in the hum-us of earth chose a veil of forgetfulness, partly to ease the deep sadness of separation from spirit and to become part of a 'blind study.' This forgetfulness encompassed the human understanding and connection to its own creation, its brothers and sisters, and all the children of its own creation. The humans descended into a place of separation and aloneness broken only by a faint glimmer of a memory glimpsed in Heartfelt moments of the awareness of the underlying force of unconditional Love in ALL things. The animals and the non- human earth beings, all of us One with you in your creation call to you to leave the cloudy confusion of your minds and join us again in the eternal knowing of Love and Wisdom held in the Heart. Spend unstructured time with us. Let us tune you again to the highest levels of vibration. We vibrate with the pulse of ALL creation and share the One breath of spirit that breathes through us ALL. We are ALL the One 'I Am' and your hope to re-connect and re-member the dream of L.I.F.E. and prove ultimately the parable that the seed of God planted in the good earth will multiply and become the fruit and flower of the sowers inspiration".

'The Herd'

The book became a beautiful compilation of some of my all-time heroes like Albert Einstein, Kahlil Gibran, and Rumi, as well as new kids on the block Anita Moorjani, and Eben Alexander who had learned by their direct experience of near-death (the microchip again), and beloved authors like Bernie Siegel. There was a host of human understanding, and then there was "The Herd." When Prima won the Nautilus award, she was the first daughter of our herd, her name was in the author position, and as I stated earlier, she was recognized as the first horse to win the award in company with Deepak Chopra, Ekhart Tole, and the Dalai Lama. The horses were gaining ground and continue to do so!

Kevin and I continue to explore our personal relationship and expand on our compatibility. Even though I would happily leave everything and follow him to a tent, while sharing our modern lives businesses and immediate space, there's always the opportunity for optimization. I'm not talking about the cap on the toothpaste, the toilet seat, the shoes and clothes littering Kevin's wake, the mess he leaves in the kitchen (although he does do the dishes), which I am resigned to take care of; but more foundational issues that have long been divided by a 'Berlin Wall.' They first raised their heads when we were first getting ready to cohabit, and I needed to coordinate the addition of five extra bodies in his home for three. Those issues are tidiness, neatness, cleanliness, but most of all, I have a diametrically opposed view on clutter, and Kevin could be called no less than a hoarder. Yes, he is a sentimentalist, but to the extreme. When you need to save every love letter from every girlfriend, and a copy of yours to her, rocks from the beach in England when you were five (and I don't mean special ones – sometimes they're simply nondescript gray specimens), a whole shelf of sweaters knit by the ex-mother-in-law with not just one or two moth holes but some resembling Swiss cheese, it's a bit much.

It was lucky for chemistry in those days, as most situations could be solved by removing his pants (the ones he was wearing in the moment). There were used Q tips and dental floss in the bathroom drawer, empty versions of every necessity stashed back in the cupboards for later use. Eventually, I came to the conclusion that there had been some damage done when his family had gone bankrupt. Kevin's father, Desmond, had grown up as the grandchild of Sir Gerald Hemmington Ryan, who basically created Life

Insurance in England in his time. The historic family home was a centerfold in Peers Lineage and also a Gilbey's Gin ad in a crusty British magazine.

Today it is a hotel that is privately booked for weddings and gatherings.

After Desmond had gone to war and returned, he invested in a family dairy farm to save him repeated induction. That was Kevin's idyllic home from birth to age seven until the family lawyer embezzled the money and threw them into bankruptcy. From there, they moved into the slums of London, and his parents went to work all day, leaving Kevin to a life of petty crime (he would steal tin soldiers in order to have a toy to play with). He also looked after the household and his younger brother Michael, until age 13 when the family immigrated to Canada and our first meeting I previously mentioned, at the office Christmas party. I suppose that scarred him pretty badly, but he has had time to get over it. When his ex ran off with the opera singer, he was awarded custody of the children and the family home for raising them in. She left the country, so he was okay, really.

Martin and Cara, the children of their marriage now firmly entrenched in our new blended family, seem to have got over the fact that their mother took the furniture and even their bedding. They are pretty minimalist compared. We have had a few opportunities to review our possessions, one in our monumental move to Kamloops, the other in our expansion to the dream house. I argue that there is no room in a 400 square foot barn suite or our current abode hardly needs the now at least 30 or 40 your old love letters, kids baseball trophies, or worse-five black garbage bags of floppy disks, DVD's or old memorabilia soaked in mouse urine, cat urine or black mold.

Our latest and long-anticipated cleaning took place recently when Kevin was going to visit Martin and Cara in Ontario for several days! I hatched my plan and gathered my accomplices; Granddaughter Rhea from Kamloops, along with our ranch help at the time, our handyman/gardener friend Joe, our horse trainer

Natalie, and Jennifer, my right-hand co-facilitator. We began by screening the office component squirreled in deep cabinetry and under things for any possible usability, and we really pared it down. We created a giant box of expired electrical cords and gadgets from old computers, televisions, cameras, and various other electronic devices. We laughed so hard, at some of our discoveries then happily chucked them! We found old slides, cassette tapes, and a wall of income tax boxes from the 1990s to at least seven years before present stashed next to the propane furnace (possibly fuel for a plan to burn the house down and collect the insurance). We worked our way from cache to cache, finding nine storage areas in all!

There were moments when we all collapsed in euphoric laughter, even recording some of our finds for a possible viral YouTube video. We had the hilarious idea to dress ourselves up as Black OPS specialists covertly sneaking into Kevin's stashes and miming out our reactions to the discoveries and what possible use they could have. There were boxes of gloves, many of which just had one side of the pair. One was even missing the middle finger which sent us into peals of laughter when I'd demonstrated its possible use - to take in the car with you for possible road rage scenarios, emphasizing the middle finger so that it stands out in the rear windshield. Really, was I enjoying this too much? We found a company called 'Lightning Trash Removal,' which boasted Safeway sized trucks, and they filled three. In retrospect, it was very reassuring as I now know it can be quick and painless (at least for me). Kevin had his nose out of joint for a while after he returned home to find everything neat, tidy, organized, clean, and hanging on hooks in their designated areas (thanks to Joe). He grieved some items that actually had disappeared by mistake, but really we left him at least five steamer trunks and a few filing cabinets of his very personal effects.

Chapter Thirteen

MORE KEVIN, THE GOOD STUFF

Now I wanted to add a section on how grateful I am for my husband (to save the marriage). In retrospect, I can see why I was so desperately in search of a husband in those years of looking for my life. I had a vision of the end result, the country life, the children, the animals, and the fulfillment of being safely held in a warm, comforting blanket of love.

When we first got together, he actually was the first man my father, who I loved and admired immensely, had ever approved of. Not only that, but he very quickly began to refer to him as 'the saint,' much to my dismay.

What did that mean? It didn't sound very complimentary to me, and I have at various times since, tried to decipher exactly what he meant. However, I think those who know me well or at least have been witness to a good lot of my exploits will agree that Kevin is always calm, kind, supportive, actually he rarely even questions my rapidly changing adventures. I believe he knows that my philosophy is 'jump and the net shall appear,' and possibly he had volunteered for the net position.

I was actually thinking the other day that he has, in fact, taken over from my father in that capacity. Throughout my life, I have always felt divinely guided, so they must be his angels here on earth. In fact, I remember Kevin's mom saying he had been marked by an angel as the mother superior told her when he was born with a lighter circle of hair right where a skullcap would be worn. It was always noticeable in his dark blond hair I think possibly he was a saint in a past life.

Sometimes Kevin will share a shaft of brilliant insight with us. In his analogy for the children about looking after our health, he explained that we start life with an empty cup and begin adding poison to it drop by drop until the cup is full and then we're done. That has so impacted all of our children that they are all very careful what they put into their bodies to this day. Another little ray the other day had me laughing at my blind spot. I'd been intent on restarting my thyroid, which had taken a long vacation. I was hoping that if I stopped the medication and really stuck to a healthy diet, it would kick in. Well, I'm a little stubborn at times, and I was running down fast. It got to a point where it felt like the energizer bunny had run out of power. I was complaining to Kevin, and he volunteered, "Well, you take the battery out of the car, and what happens?" I went back to my thyroid medication and felt much better, almost immediately.

I mostly motor on in my comfortable little rhythms with an underlying feeling of happiness and gratitude, but foundationally there is a fire and a passion connected to what I hold dear. I've been known to fight for it, particularly in my youth (thank heavens I have found wisdom and philosophy in my later years). The few times I've incited Kevin to violence (because I wanted to see what it looked like), he fought back with one arm tied behind his back or simply restrained me until I settled down. He is not exploring sainthood in this life but is certainly appreciated by all those in his office for his always calm, centered approach to stressors, and in the family, he is counsel for all. They don't ask me as my "it is all always perfect" philosophy is rarely welcomed as a tool for those who enjoy reason.

One side to Kevin that he has not fully explored is his writing ability. He is always written me long letters and poems, and his family Christmas letter is always much appreciated. He is really, funny, and often a bit Cohen-esque in his obscurity. I will often ask him to translate. I want to include what I think is my all-time favorite of his lighter writings so far.

His historic gifting of cacti to me had always been cause for reflection, much like my father's gift of 'the saint,' and he managed to turn it around to be seen in another light:

A Valentines Floral Episode.

"Madam, he said with a keen, enthusiastic enquiry, where might I ask, are your really big and beautiful cacti?

The transparent salesperson smile faded with first a, 'did I hear this man correctly?', to an incredulous impatient disbelief, to the mirth of the insensitive at witness of a simple-minded idiot, and back again to the plastic array of sales glitter upon her facial features; all passing as in the smooth movement of an eclipse.

My dear man (with a patronizing sneer) she replied: "This is Valentines'. Why on earth would you want such a prickly, ugly shrub on this day of soft roses and sweet delights?"

"Well," he said, eager to engage in defense of his earnest request, "the cacti are the ultimate Valentines' gift. They embody all that is uniquely special in a woman. They are firm and resilient, well-rooted in the earth, surviving on sensible balance. Yet these rounded forms and moist, succulent interiors are not for just anyone's touch. The prickly protection ensures only the most suitable and serious encounters.

"My, my, you do see things a little differently, sir." "What of the red rose, most normal persons preferred choice?"

"They are but fleeting moments of beauty. Soft and vulnerable, yes, but whimsical and soon to wilt. Would this be the woman of your choice, a flash of beauty soon to be discarded? Unless you replace them regularly, of course".

"And what of chocolates? Surely they honour a woman's soft centre and sweet demure?"

Hardly," he replied, "How do you feel after half a dozen? Besides, to most, they are a painful treat full of the anxiety and hesitation of the forbidden".

"And what dare I ask, do you recommend, sir?"

Obviously, the natural nectar of natures' fruits. There you find seductive shapes and satisfying sweetness.

"Well, whatever," her pasted expression peeling at the edges, "I still feel flowers are my choice."

With a final burst of explanation driven on as only the true believers are, he presented his crowning illustration for the confused soul before him. "The flower," he started "of the cacti, is not predictable or common. The flower of the cacti is very particular; it shows its beauty only when the earth dances the right tune, a perfect conjunction of light, warmth, and earthen abundance. And even thus there is no guarantee, for the soul of the plant when feeling just so and only thus, will burst forth in a torrent of beauty not just vibrant and vivid in colour, but with the intensity of commitment, the luminous luster of orgasmic abandon and sensual serenity for a perfect performance orchestrated only by God".

"That's quite enough, sir. I would ask you to settle down, regain your breath, and no, we don't have any cacti, please leave".

Deflated, depressed, and forlorn, he withdrew silently from the store. His continued search saw only repeated episodes of the first. Empty-handed, he returned home without a gift for his Valentine, yet she the greatest gift of all.

It was small recompense that he noticed weeks later the once barren florist now full of cacti big and small shaped in wondrous forms, all waiting for the moment to express their own particular 'gift from God.'

I did publish some of his poems in 'Sabbatical,' my fourth book, along with a couple of my own. I still see Kevin slowing down (I would never say retiring) enough to do some writing in his twilight years. There is a lot of wisdom there peeking out from its cloaking. I'm looking forward to those times when we can sit on the porch of our last hurrah, our small comfortable one floor seniors home overlooking our little 'Herd Lake,' the horses' (and Jennifer's) swimming pool. I can easily go forward to that time and see it all clearly. Kevin and I will prehumously settle our affairs, finding the perfect couple who share our dream on this sacred land with its animal higher beings. At one time I had wanted to call it Eden or Heaven or Paradise until I was directed to 'Gateway' as it has become for so many (it was for a time nicknamed 'Gatewhirl' by Kevin for its revolving door aspect on those whose energies are either not in sync or have reached their saturation point). We will stay on in our retirement home in an advisory capacity until our demise (note I didn't say infirmment as we plan to leave this plane promptly when our contributions expire).

Being my life partner, he's helped mop up the spilled milk and the vomit dripping onto the bunk below. He's been there to dry my tears and celebrate my successes. There are so many shared memories in a lifetime together and those yet to come. He's the cat janitor, and I'm the dog. Strangely he has steered pretty clear of the horses although he is on call for forking down a large bale of hay or chasing the herd from a break-in to the front yard back to their area in the back. He's particularly good at that as the horses are very respectful of him as they are not sure exactly what he's capable of and partly because he has amazing focused energy. But it always confounds me that he can be intensely fixing the waterer while a crowd of curious horses gathers around him gingerly poking or prodding with their noses, and I have to call out the window "say hi" or "reach out and touch someone." Is he jealous of my affection? Interestingly he has never shared my love of animals being raised with only a family dog (who was not his favorite choice). He does like cats, though; maybe he is attracted to their independence, which is interesting as that is not my prevailing character (or maybe in many ways it is).

I still prefer his daily presence in my life and, in fact, am a bit lost without his grounding influence. In the area of grounding, we have always had a transcendent (probably from many past and between lives) affinity and transformational sex. It can't even be relegated to 'sex' although that has been monumental. It is more a melding. I have tried to transcribe those feelings over the years in cards and illustrations for cards. The ongoing, pervasive quality is a complete loss of space, time, and dimension and a powerful body experience in a bodiless state as it is from beyond time and space and tantamount to all experience (except possibly a God experience). It is sad that frequency is declining with my hormones as I was all over him in my years of possible conception. Sadly, my interest level has faded with the blush in my cheeks, and although it is still beyond satisfying when he initiates it, I have lost the urge to pull his pants off. I still look when they're missing and admire his naked torso, but it has faded to a warm, cozy feeling rather than intense desire. It is somewhat like my attraction to my children; I have lost the desire to pull them into my arms and celebrate their softness and innocence and now admire them from afar.

Strangely I still find fur and four legs adorable, irresistible, and preferable to pink skin and strange looking appendages. I have always looked at humans as odd creatures, far less attractive than most animals. I mean really the holes in our faces are so apparent, and our hair is relegated to odd places and a big crop on top of our heads, never mind men's strange facial growth, now so fashionable and even more unattractive at least to me. We, women, spend hours changing our hair by adjusting the length or spinning it in one direction or another or tying it in knots. And what about our nails! Claw-like but totally useless for any purpose. We embellish them with now intricate designs and myriad colour combinations, not even matching. And what about piercings! Who wants to get snot stuck in their nose ring or floss their tongue bar? And then there's tattoos! Who wants to be black and blue and read all over? I think it's because we secretly all agree we're not the most attractive species, and we keep trying to make renovations. I wonder when there will be glue on fur? I'll be all over that!

I remember when I was in my teen years and introspective about my choice of body, looking at my hands and feet in particular and thinking, "this is not me" they look weird and alien. And what about aging? I can look kindly on the ravages of time on some others, particularly those whose spirit shines clearly through the casing, but on one's self, it is more difficult to overlook the sags and bags of older skin. Interestingly it doesn't bother me on elephants, although I have never found skin to be anywhere close to fur in its appeal. Fur is forgiving, silky, soft, comforting like a cozy blanket, and just has always called to me like a familiar friend or almost a memory. I'm sure I have experienced many animal lifetimes (not even lifetime's as I believe we, being all expression for all of time, dimension, and realities, can experience any aspect of creation, being creation itself) but it is still a mystery why I chose to be a human.

Sleep is another mystery that is always nagged at the edges of my understanding. Why do we put ourselves back in the toy box every night after playtime each day? And who is playing? Are we the players or the toys? Most likely, both as there is only one of us in the room.

Chapter Fourteen

THE LORD IS MY SHEPHERD
(OR; CHRIST AND 2020 RETREATS)

"**A**nd God so loved the world that he sent his only son..." Who of all creation is not the child of God?

I do believe in Jesus Christ and the Christ story and the fact it represents the possibility that we can all attain Christhood or Spiritual perfection while in a material body. We are all immortal beings of light and energy and intention, born of great love.

I have a long relationship with Christ and a very real experience of being with him in a Mary Magdalene experience. I do not believe that I necessarily had an incarnation as Mary Magdalene, although it is possible. I believe we can experience any aspect of existence that draws us, not necessarily for a whole lifetime. I do totally love Christ, the man, the image, and the possibility, and he is very real for me and always has been for as long as I can remember. We have always had a relationship, and I have memories of conversations and meldings, microchips of knowing that have been with me always from my earliest memories of sitting together in my childhood at the bottom of the stairs with the closed door and the small grill of light as my only connection to the adult world below. I always felt his love and knew I had come here to do something special for God and Christ as God, and the nudging was another one of those mysterious wonderings that was just at the periphery of my understanding. One of the exercises we do in the retreats is search the numerology of our birth names and birth path (or date of birth.) My birth path is number 33, 'The Christ Path' and my name Ann Elizabeth is 'Grace,' 'Oath of God' As they say it is all written.

Christmas about eight years ago, I was communing with him on his birthday, and I suddenly had the realization that I would paint his portrait. I had known this for some years, but I knew it would have to wait until I saw him clearly. It was time. I went to work in January that year, and when it was finished, he was radiant, beautiful, and his hands glowed. People began to say they experienced healing just from looking at his image, which I had shared online. I was told that he needed to be shared with whoever asked. He sat in my home, and it felt a little strange as there really wasn't a place to do him justice. We were far from being a religious institution.

My friend Connie (from the Equinisity film and our first retreat) offered a solution. She suggested he go to an interdenominational religious affiliation between a catholic priest, Father William Treacy, who is now 100 and co-founder of Camp Brotherhood, with Rabbi Raphael Levine at Lake Mc Murray outside of Mount Vernon. They were best friends and bought and restored an old dairy farm into an international interfaith center for peace and brotherhood. It is now Camp Korey, a place for children with life-threatening diseases to find support and joy with their families.

Father Treacy still lives on the land and does his walking prayer and meditation in the woods daily.

They were happy to have him, so we met at the border in a joint parking lot of US and Canadian citizens, did the shuffle, and luckily no one said a word (probably because Christ was in charge). He lived there until it was closed, and then Connie came to the rescue again and moved him to her Catholic church (Connie loves Christ as much as I do). Again, he lived there for some years until I was directed that it was time for his return.

This was after the Healing Temple with its pyramid roof of crystals, and stained glass had been erected in 2019. The healing temple was inspired by Dolores Cannon, a famous hypnotherapist that discovered how to lead people to experience past lives and 'Between Lives' (The title of one of her 17 books). Her subjects described the 'healing' temple on 'the other side' (along with Akashic records, the tapestry, the life review, Etc.). Our version had a beautiful glass pyramid roof of multi-colored healing crystals like the one on the other side, which I wanted to recreate in a small way.

In following Dolores, her books, and YouTube videos all one winter, I decided I wanted to book a session for past life, between life exploration partly to try to uncover why my knees were getting progressively more painful and problematic. I had some past life recall before, and in several of my lives, I had died from a complication involving my knees. One life I had in Wyoming with Kevin, who was a guide and trapper spending a lot of time in the forests there. He was caucasian, and I was native again; also, an artist who painted native portraits and my horse in that life was Diva in this life. When he didn't return from one of his forays, I got on Diva and went off in pursuit. I have a clear image of being in a valley bottom with hills around and a hunting party shooting us, me below the knee (which I have a circular birthmark the size of an arrow to commemorate) and Diva, a bullet in her hip.

We both died together in that life, and it was my brother in this life that shot me. Strangely my brother and Kevin have never been friends. Interestingly, I was rafting down the Rio Grande river when I was given that vision, so when I returned home, I checked Diva for a birthmark too and found a jagged lighter patch on her hip where the bullet had been that I had never noticed before.

Anyway, back to the Dolores Cannon work, I found a hypnotherapist in Kamloops that had taken her course (amazingly) and booked a session. It was an all-day event. She came to the ranch, and I was induced after an interview and explanation of the process, fully expecting to go backward in time and understand about my knees. Instead, I stayed on the spiritual plane (between lives, I guess) and went to an amazing healing place that was a white room with a big healing table surrounded by ascended masters. They were all wearing beautiful colored rings of different jewels; Sapphire, Emerald, Topaz, all different colors, and their eyes shot beams of multicolored light. They were gathered around me and had their hands on my body, sending healing. To one side against the wall just out of my vision and slightly behind my shoulder sat an enormous presence in size as well as emanation, and he came over and took his place at my head so that I couldn't see him, but I knew it was Christ. He put his huge hands over mind, which were one over the other on my chest. I can still feel them and their size overlapping mine. He took off his Garnet ring (which I knew he wore from the portrait and my past experience) and put it on my right ring finger, which I discovered later was the one they used for marriage in early Hebrew ceremony. I was told that I had a portal to visit there whenever I wanted.

The session ended with Christ telling me that it was time for his return. The healing temple was ready, and he was to be on the wall there. He showed me the Gateway 2 Ranch sign on our entry gate with chains hanging a sign below the large Gateway 2 one "Christ was born here," and more chains and another "Home of Christ," and I laughed out loud and of course didn't add the signs. I guess in a way he was, and we now had a place to hang his portrait. After that amazing session, I felt sure my knees would be healed and did experience a brief period of less pain and more mobility, but it was not the miraculous healing I had expected. I believe there are many factors on the path we are guided along, and always some reason that we experience and understand in retrospect. That can mean, years later.

I was at my friend Trudi's telling her of my experience soon after the event, and I could still feel Christ's ring on my finger and his large warm hands over mine when I would go to sleep at night. Trudi, ever the skeptic, looked over at my hand, and jokingly said: "I can see it, I can see the ring." Briefly, I thought she could, being ever the believer as one of my primary character traits. She laughed, but it planted the seed that I should look for a ring just like the one he gave me to cement the experience. It was a large square Garnet, so I put out a request to find it. One day while driving downtown close to the Golden Buddha, a metaphysical shop in Kamloops, I was prompted that if I found a parking space in front, I should stop and look. There was an empty 15-minute parking zone directly in front, so I parked, ran in, and found it immediately. It was perfect and has been on my finger since.

Christ was returned by Connie again when she came to an August retreat and was happily ensconced until September when our retreat came with 'the three Witches' as we dubbed them, one a white witch, one a black witch, and one that admittedly practiced Wicca. That evening we experienced the biggest lightning storm ever in 20 years centered around the Healing Temple, where we had gathered to share the talking stick and our day and were watched over by the portrait of Christ. I also had new ranch help (who in retrospect was pretty witchy) in the kitchen preparing dinner.

The lightning flashed all around us, and with the immediate thunder, we knew it was directly overhead. We could see it strike the fence nearby, dancing in blue arcs along the top rail to the nearby electric pole and causing a huge explosion that took out our power. Then to many surrounding trees again and again. Thank God (no, Christ) that there wasn't a fire. I felt in awe and completely safe at Christ's feet, but it was definitely strange, and we were out of power for the following three days. There were a lot of threes involved, and the whole retreat was the strangest ever. I heaved a sigh of relief when our coven returned home, and we could go back to our normal happy life here.

As always happens at retreats, people find each other and connect, who are dealing with similar issues. Each unique group is met by a different version of the Herd, and by looking at that reaction, we can tell what some of those issues are. There are those who are working on opening their hearts, connecting to oneness, healing, taking a step forward, ones who want to explore past incarnations; sometimes, ones lived here. I had looked at my Collette Baron Reid cards "Wisdom of the Oracle" this January to see what the coming year would bring and in a quick overview it said:

"You begin to attract the perfect people to support your dreams. Mentors, business partners, helpers, employees, creative partners, friends, and strangers open doors for you and step through the portal to join you in a harmonic dance of collaboration, commitment, and co-creation. This is what you've been waiting for!"

I should share the story of 'Herd,' our most recent award-winning film. When Lisa Lightbourne filmed 'One With the Herd' and we won Best Documentary for it, we decided to go to the Equus Film Awards in New York. While we were there, we met Stefan Morel, who was there to receive an award for his film 'Blindspot' We all hit it off immediately and talked about horses and films non-stop. I also met the 'Horse Lifestyle T.V.' people who later offered to come to a retreat and film our next venture. We later discovered that conception did not have the horses' full approval, and it was canceled last minute. As always happens when spirit is doing the directing the project became 'And Now, For Something Completely Different' and I wound up hiring Stefan instead who brilliantly executed 'Herd' Our many award-winning film. Another trip to New York ensued and a great chance to hang out with Stefan and his wife Pam (and a carriage horse in Central Park)

Liz with her sweetheart in Central Park

Connie connecting with The Herd

Jennifer with Amora and Omi

Sonja with The Herd

Our Award-Winning film "Herd"

The Horses in a Healing Circle at the Tent Village

Lizzy, Jan and the Herd

Liz's 60th with Rhea, Denny and Izzy and spirit friends

Tesoro and llama help heal

Horses and Spirit Friends hold space after a Healing

Liz with portrait of Christ in the Healing Temple

Herd using Sacred Geometry in the barn

Chapter Fifteen

HORSE STORIES

As I am writing, the herd is all gathering outside my window, which always means they have something to share. They would do that through the previous five books and whenever I would have an interview. They've reminded me to tell some more horse stories. I have shared many in the other books, but there are always more. Horses, and all animals, have a very developed ability to transcend time, space, and substance.

My Herd, once connected to someone, will continue to visit and counsel those who welcome their input. Just the fact that those people have been given the microchip and access to One Mind is readily apparent to other horses (and animals) who can easily connect to our minds. There are stories of some of our students in horse therapy careers in which the horses were previously underappreciated and burnt out returning home to discover their horses had picked up on their new understanding and thereafter co-operated full force. And others speak of an ongoing communication whereby they are connected to horse wisdom which is an unlimited resource of universal truth, far larger than just human. They share abilities of the clairs (clairvoyance, sentience, etc.) and connection to universal wisdom. I would like to share Avis' information that a therapy group told her:

"After I returned from 2019 Equinisity Retreat, I visited a favorite equine rescue sanctuary. I was still floating from my time with The Herd, so I came loaded with questions for these horses. First, I asked if they 'knew' the horses in The Herd. The reply was that they were in touch with all horses on the planet... and off. But they hadn't specifically known The Herd members until they met them through my energy field. They seemed intrigued and grateful.

My second question was: you live here in comparatively small paddocks, compared with the hundreds of acres on which The Herd can run/gallop/fly. I can imagine how important that is to Horse. How do you deal with that? Their reply was simple: "We gallop and run and fly on the Interplane, Avis, as you well know because we take you with us when you ask." Aaah - true enough: I love riding bareback on a galloping steed on the Interplane. For starters, I stay on! And we often do end up flying into other dimensions.

My third question was: In such a caring place, yet still confined more than is natural for you, do you consider yourselves 'lucky' or 'unlucky'? (OK, I knew it was a silly human-question but...) Their reply was: "Both/and. All and each of us are part of the experiment of re-creating peace and harmony from the chaos of seeming division. We all choose our roles to play and, therefore, our locations. One of our goals is to help humans learn to honor and hold sacred all Beings, indeed everything. To work directly with humans on this mission - especially as Horse - is to be subjected to dishonor and being 'used.' This is to be expected, as humans (as a whole) are currently in the early stages of development in the healing of The Myth of Separation, so they are very self-centered and hold *no one* else as 'equal' or 'to be respected.' Often, not even themselves.

As the Unity Quotient grows, some humans 'rescue' some animals, and both get to experience 'respect' and 'holding sacred.' This Sanctuary space/place is very peaceful because so much is 'held sacred.' We Horses don't really quantify experiences as much as you humans because we know our purpose, and we are in touch always with The All.

That opened a door for one of the dilemmas that interests me most on Planet Earth. "So," I proclaimed, "if EVERYTHING is considered sacred - what about eating?" Their answer astounded me: "Humans have very little experience with the level of consciousness that Nature experiences on Planet Earth in this regard. You think of "eating something" as selfishly killing it (if you think at all). You are little aware of the releasing of your form for the betterment of another and The All. The only way that you "give up your life for another or the good of the whole" in any sacred way (typically) is when you go to war. And then the 'killer' does not hold 'the killed' as sacred but rather as an enemy."

The horses also have no attachment to one particular lifetime, and will often be reborn here again if their mission is not yet complete. Many, when they call in The Herd in meditation, will see members that are deceased and those they have never met. One story is of a retreater who, in one of our led visualizations, saw a foal standing by L.E. and asked who it was. It was indy, L.E.'s daughter, prior to Crystal, who was a complete twin to Crystal in markings.

The story is that for a period of time, several of our beloved animals were sacrificing their lives for the development of their abilities to enlarge hearts that debuted at the retreats. Ra, our Leonberger dog died at age seven. After receiving her breakfast cookies, she walked out onto the porch, lay down tongue hanging, stone dead. Athena, our whippet, suddenly developed heart palpitations (after a few years of running flat out without any sign of difficulty) and died of an enlarged heart at less than four years old. The same thing happened to Hami, her brother. Theo, our Doberman at five years old, overnight developed symptoms, and an enlarged heart and was dead days later.

Indy, L.E.'s daughter (these all happened around the same period) was born beautiful, vet checked and declared perfect, and three days later, she wasn't nursing, milk all over her nose, so we rushed her to the vet. They didn't even check her heart as she had been declared healthy on the vet check days before. They tested everything else all day long as I kept protesting, "can you please check her heart?" which they continued to ignore. By the end of the day, they had run out of ideas and decided to listen. Guess what? She had an immensely large heart, and they had to put her down. Poor L.E. was desolate and disbelieving, and we had to bring the baby home with her until she could come to terms with the sacrifice. A year later, on the anniversary of Indy's death, she started running around calling, smelling places where they had been together, and then a week later gave birth to Crystal, an exact duplicate of Indy but with a perfect heart.

Then we had Luxy and Monet, our two most Angelic horses. Luxy was born an angel, a beautiful angel with four perfect socks and a double star on her forehead. I suppose that brought her to the immediate notice of my friend April who fell in love and wanted Luxy as her first horse (a dream she had had since childhood). I usually don't sell Herd members, but April promised to keep her here, and April was my other best friend (as well as Trudi). April and Luxy learned everything together which is always discouraged, but Luxy, being an angel, stood quietly, did whenever she was asked and took April through years of riding in the hills, Parelli training, and the privilege of being most beloved by an animal being (it went both ways). April took Luxy to Vancouver for a year to have her closer and show her to April's mother and all her Vancouver friends, and there was great celebration at her return. April driving up in her truck, was just rounding the corner by the riding ring with the window down in the horse trailer and Luxy's nose through the grill. All of a sudden, there was a unanimous call from the herd as they had been lined up along the paddock fence for probably the time it took April to drive from the highway up the hill! Just like the book 'Animals That Know When Their Owners Are Coming Home' by Rupert Sheldrake (proof of entanglement). April stopped and let her out, and she ran the distance of the pasture to the welcoming committee. Prima and Epona, who loved Luxy best, licked her all over her body for 5 minutes. We were very teary-eyed in witness.

Anyway, I was explaining earlier that the Herd is always together even in their spirit bodies. They are reborn into the Herd and die for the good of the Herd. Luxy died at age 13, which is far too young for a warmblood, and she sacrificed her life for my firstborn granddaughter Kaia. The story is that morning, Luxy came in with the Herd and began falling on the ground, just dropping and lying there for a few moments, then getting up. We watched her do this several times and then called the vet. The vet did a colic check (it was the weirdest and most severe presentation of colic I had ever seen or heard of). The vet couldn't really find much, tubed her, gave her some pain med's and left.

Meanwhile, Kaia had driven her brother down to the guesthouse with a lawnmower in the back of the Kubota to mow the grass. On her way back, as she was driving around a sharp corner, Thaelo began calling and chasing the Kubota to tell her something. She went up the bank, the Kubota tipped on its side, and luckily she was thrown across the road to safety as the Kubota would have crushed her. She later shared that Luxy had been foremost in her thoughts at the time for some unexplained reason. I was not witness to the Kubota incident, and we got a call that it was in flames, black smoke streaming out, and April went to deal with it in fact, saving the day and the forest from burning.

Meanwhile, I stayed watching Luxy, who at about the exact time of the Kubota incident, fell to the ground again, and this time, she wasn't getting up and was in considerable pain. I called the vet and April, and we watched her die. It was awful, we were helpless, but we were both aware that she had traded her life for Kaia's. She was also Kaia's favorite horse and in her angel capacity took care of Kaia on her first rides, and always kept her safe.

Weeks or maybe months later, we were out in the third eye chakra on the land, and the Herd came galloping through following a spirit horse, Luxy, in great excitement. I believe she has come back as Flower, our new grey pony. Flower, like our other pony Merlin, was not born here, but she called to us. The first-year Jennifer was here, we both saw a grey pony calling to us. She showed herself clearly and said she needed to come here. I went to the online ads, and there she was. We hooked up the horse trailer, Jennifer driving while I prepared a spring green bucket with the name Flower on it and off we went. She was darling, so we got out her bucket and loaded her in the trailer, her owner, much impressed at our certainty.

Prima, Luxy's dearest horse friend, immediately took her under her wing and shielded her from the curiosity of the Herd. She was untrained, but came knowing all of our Natural Horse Friendship activities, a diminutive version of Luxy, and has fit right in playing with the foals as Luxy once did, and never putting a hoof out of line. Prima remains her best friend.

We have our difficult birth stories. Epona just about lost Paschar (another one of our angels) and her life as well. About a month before Paschar was due, Epona showed up with difficulty breathing, and a bloody and pus-filled discharge from her nose. Off we went in the trailer to an expert in Vancouver. She had a lung abscess, and they wanted to abort the baby. What?! Abort a ten-month-old fetus? Epona and I both said, "over our dead bodies"! So, she hung in there, drugged, losing weight, and seriously ill for two weeks until she took the turn for the better. They wanted to keep her there for the birth, but we were out of there, driving home to Paschar being born the next day. He was perfect, although three weeks premature, and we welcomed him to the world and made sure he was nursing and went to bed.

The next day Epona, who is always funny and easygoing, wouldn't let anyone near her baby, even trying to chase me away once until I reminded her that I was her mom. Paschar, who had been very quiet right after birth, tried to chase everyone away as well. Imagine a tiny foal rearing and running at you. Their issue was that they both knew that Paschar was threatened by abortion, and they were fighting for his life. It took some gentle words, and gentle handling, Kaia, in particular, would go up and hold his head in her lap and talk to him. We would tell him he was our angel (Paschar is the angel of vision), and he was safe and loved. To this day, I look into his eye and remind him he is my angel, and his eye gets all soft and reflects his gentle and sensitive nature.

Diva almost died while giving birth to Serene. When her water broke, I reached in to find not two hooves and a nose as you would expect, but a mysterious hairy appendage that I couldn't identify as mane or tail. "Emergency call to the vet, get the mare up and walking," Phyllis Loose says in our bible for birthing, 'Blessed are the Brood Mares.' We did as suggested, and 45 minutes later, the vet arrived, attached chains, and the fight was on. When they finally pulled Serene out, she wasn't breathing, and they were resigned to having lost her. We were definitely not going to accept that and went to work, April giving her mouth to nose resuscitation, and Serene took her first breath. Diva was so heavily drugged, but there was a darling picture of their first meeting when "baby spider legs" as Peter called her first stood up, and they first met. I love the soft nicker characteristic of mommy-baby talk. Noses touching, they stood there, blinking with the caption "what just happened?" hanging in the air.

Ellie had a difficult birth and was in the birth canal for far too long, making Jennifer and I a little anxious. Her shoulders were stuck (she was going to be carrying a huge weight on her shoulders being L.E. reborn and was maybe a bit reticent). She did make it eventually, although her nose and mouth and tongue were squished quite badly, and it took hours and lots of help to nurse. Later on, in her early weeks, she got left in the corner when the Herd exited the small gate. She panicked and tried to jump the fence, hurting her shoulder again. Those expressions "take a step forward" (my knees) and "weight on her shoulders" (Ellie) always tell the story.

Then on the lighter side, there was Epona who got her nose out, and began interacting with me trying to pull herself out with her two forelegs (almost doing the breaststroke) and Prima (who has zero tolerance for pain) leaving the birthing stall and running circular laps around the playground every time she had a contraction then running back and having the easiest birth ever with Picasso to a crowd of witnesses who had gathered wondering just what she was doing.

Chapter Sixteen

A FEW MORE HORSE STORIES

Onward from birthing, we have our Winnie dying story. Winnie would get bouts of laminitis due to her insulin resistance, usually in the late spring. She had done really well that year, and I began letting her out for longer periods. She came back one day lame, so I locked her in, but it was taking a while to abscess, which it always did, appearing out of the coronet band, and then she was on the mend pressure off. Well, this time, the vet was already over dealing with another issue, so I asked him to take a look and, in the typical "I'm the vet I know better" manner, he told me to step aside, and he dug out Winnie's front foot (really dug it out). I kept protesting about the coronet route she always presented, to no avail. Well, that was it. Winnie, who has not much more tolerance for pain than Prima, lay down to die. She really scared us and wouldn't get up for a week. I called another vet who gave us painkillers, which seemed to help, but she was down most of the time, and they were giving up.

I called Lizzie, my student, and mentor who runs 'Whole Horse Consulting', and is a vet tech as well as homeopathic herbal expert, and also an animal communicator and intuitive. She usually helps me from a distance as she lives in Texas. Lizzie <u>always</u> solves our issues with her alternative bag of tricks. This time Lizzie said I think she is ready to go. "What!?" I said, "not you, Lizzie, that's really scary." It was the last day of the retreat, Singing Day, but I really didn't feel like singing, and all the participants were in the big barn listening to my Rodgers and Hammerstein & favorites soundtrack, all worried sick and doing their best to visualize a healing. I went into the birthing stall, devastated, lay down spooning Winnie, and whispering "live" actually pretty loudly, begging her to live. "Live Winnie, live. You will live Winnie!" we got into the part of the track with 'How Great Thou Art,' and finally, 'Amazing Grace' and Winnie rose to Amazing Grace, and the crowd went wild. Hugging, crying, and declaring it a miracle. She was healed! Oh, my God! Last season she had the odd slight lameness but no biggie. Wow!

Having mentioned Lizzie in the Winnie episode, I have to tell the story of Crystal losing her first baby to a miscarriage. She came in when the baby was about a six or seven months old foetus with bits hanging from her, and we were so disappointed but anxious to help her clear. I called the vet as it wasn't clearing easily, and I was worried about infection, and they gave her Oxytocin, did what they could, and left saying that the next morning they would cut it out. "NO!" anyway, the next morning came, and the vet was on his way, and I called Lizzie. Lizzie did a distance check-in, told me to go to the barn and focus orange healing light on her sacral chakra, and give her Pulsatilla a homeopathic remedy every 15 minutes.

Well, one application worked, and the baby cleared immediately, and the vet arrived to me gushing about homeopathics, and he simply rolled his eyeballs (he was the same guy who treated Winnie) Poor Crystal. She didn't even look and stood stoically facing away and then walked off, not looking back. There is a story to the fetus who was a lethal white pinto (even though that was impossible as a very specific combination of genes is required which the sire didn't have). We were given that it was Monet coming back for the brief time he needed, and Crystal volunteered (being his half-sister and they were born at the same time).

Monet was Winnie's, first baby. I told of the birth in another book, and he was another angel. Our only pinto at the time, he would follow us around like a puppy. He loved everyone and didn't return with the Herd one February morning. Winnie was desperately calling and calling to me to follow her, and she took me there. He was dead of colic, the snow melted around him, and hundreds of hoof prints in a circle around him. That phrase about 'only the good die young' was true. He was my favorite. He was everyone's favorite. Crystal, Miro, and Serene were born with him that spring. Monet came back in spirit form in a dream and bred Epona to make Paschar. He had something to finish or help Crystal with, and he returned for that brief period in-utero.

Winnie went on to birth Prospera. Actually, that was a story! Winnie was always very interested in sex and educated all of our young stallions in the facts of life (at least until the vet was called, the morning after pill administered, and the gelding done). For Prospera, Winnie was allowed to have indiscriminate sex with a gorgeous Andalusian stallion resulting in my first and only Iberian warm blood. The story is that we trailered her to nearby Kelowna for a week of fun. I was there with a large degree of trepidation, having heard many stallion mare stories and only having my Chako stallion experience to compare. We arrived, and Winnie was presented over the fence to prince charming (actually Padrino, Paul Dufresne's stallion) that was bred by my best friends Bette Lynn and Albert. They had suggested for years that I should do it. It took about 30 seconds for them to start kissing; it was love at first sight. Paul, being very relaxed about his stallion and allowing him a life of freedom and friends, opened the gate. Winnie ran in and presented herself, and that was it.

Not just it, he was such a gentleman (Padrino I mean), and afterward, he leaned forward, kissing Winnie's neck and telling her how wonderful it had been for him. Then a while later, they did it again, and then again. Paul served food and put out a pile of hay to distract them. Winnie always in charge, claimed it, and Paul put out another, and another and Winnie kept claiming them obviously ravenous afterward. Finally, she picked one Padrino could have. She stayed for a week, and it was honeymoon the entire time. They had so much fun that the follicle was washed away, and the next cycle, we bred her with artificial insemination and shipped semen from Padrino. I felt so bad taking Winnie back. Not for her as she would have all the future yearlings at home, but for Padrino, who called and called after her!

The horses are out on the land every afternoon in grass season and return every morning at dawn to wait to be let into the paddock and their healing work. Even between retreats, they go into their barn church (Connie put a cross on the wall with a picture of L.E. and Crystal to bring attention to the fact that it is their church). They also have sacred spaces on the land and can be found in ceremony gathered in their healing circles.

They are always welcoming to humans who were called to join them. Often when we're in the Spirit Lodge, they will grace us with a visit or drumming will call to them. One night when Mark Mottershead (of Horse Conscious) was here, we were having our sacred fire at the outside fire pit by the tent village. We could hear their hoofbeats, and we walked to the nearby valley to see them all in a line to say goodbye. Tesoro the steer was with them, and as the herd came over running down the near hill, Mark said I can't believe he can keep up (Tesoro was already his humongous self). Well, Tesoro heard him, and he began leaping and kicking out to one side as he did in his YouTube video when he was younger. The video title is 'The Steer Whisperer' if you want to look it up.

He ran down the hill, up the one with our Spirit Lodge and took on the solar panel in a typical Don Quixote fashion doing battle with the phantom intruder, knocking it over and then pleased with himself, casually looked over to Mark like "you saw that didn't you"? Tesoro was yet another angel and always looked after the people who were a bit nervous about the horses. He was like a giant beanbag couch, kind, gentle, always good-humored, and in love with Crystal. He used to lick her all over when asked, and once or twice, to my surprise, actually mounted her (she was just fine with it) and attempted to make a 'cowhorse.'

The horses regularly sleep in the meadow next to the tent village and sometimes circle the tents to help with additional healing needed. That is why I do encourage people to stay there. That, and the interface with nature. The area has owls, deer, coyotes, the very rare bear sighting, and magical trees, fairy circles, tree circles, and double and triple trees (talk about forest bathing, it is forest immersion).

The other thing the horses are sharing is riding stories. Although we don't do much anymore as it seems, the human/horse relationship is morphing, but I will never forget my amazing relationship with Prima, my firstborn who I learned all my horse riding, Parelli, Natural Horsefriendship, and liberty play with. We had such a powerful connection that I could gallop bareback in the hills with just a halter, riding her with nothing in the playground just asking her to climb up on tires or come over and let me mount. I did ten years bareback and bridleless, and although I didn't even ride last year with my knees, every time I hop on no matter how long it has been, we are one yet again. I will hold those memories as some of the biggest thrills in my life. At retreats now, we rarely have people who even want to ride, but if we have an experienced rider who wants to feel what riding in a halter and a treeless saddle is like, Miro or Prima will help. We don't believe in forcing the relationship or actually a ride with no relationship first. As I say in liberty play, which the horses are happy to teach first, you have to date, have fun, like each other, and maybe when you have a relationship, go to 'third base.' I think the concept of "kick 'em to go and pull 'em to stop" is becoming a thing of the past.

The last horse story is a conception story. Crystal, after her miscarriage, had an ongoing level of infection, and when we tried to breed her the next year, the vet was very skeptical that she would be able to conceive. We flushed her with various antibiotics and my suggestions of alternatives from ionic silver, grapefruit extract, and a special enzyme blend from Australia designed to dissolve the antibiotic-resistant film secreted by the bacteria. We made several attempts, gave her some time, and then I talked the vet into breeding one more time. It was so important to me that Crystal have a baby as her mom was gone, her firstborn died, and she was processing what seemed like grief. She was bred.

The day the vet was scheduled to come and ultrasound confirming the pregnancy, there was a retreat going on. We shared with the group our expectations and fears. The horses were doing their morning meditation some in the main barn where we did the ultrasound, and some in the birthing stall. People were in the central seats with horses standing over them.

All of a sudden, Paschar (our angel of vision) came over in a big hurry from the birthing stall; his eyes rolled back so you could see the whites, and began searching the barn calling in the characteristic nicker of baby talk. He was following a spirit baby all around the barn. The other mothers were getting upset with him thinking he was after their babies, but he was definitely after one we couldn't see. For whatever reason, he then ran back to the birthing stall and bred Magic (something that he had not done since his pre-gelding days). The vet arrived and announced the miracle that Crystal was, in fact, confirmed in foal. We think Paschar was seeing the spirit of Ellie entering the fetus as the reincarnation of her grandmother and beloved lead mare L.E.

Chapter Seventeen

LAND STORIES AND 'LIFE' STORIES

The land here at Gateway tells its own stories. Over the years, we have uncovered many of its mysteries and have been privileged with its wisdom. Scientists are beginning to realize that Earth or Gaia is the larger body of consciousness, and beings are dependent upon that body, a diverse macrobiome. My theory is that we're all consciousness-expanding itself, fractals of God ever replicating, from the universe to the stars, planets, planetary life, to microbial life. I remember getting my first autograph book and my dad writing in it "Big bugs have smaller bugs upon their backs to bite em. Smaller bugs have smaller bugs, and so on ad infinitum".

Huh?

I get it now. As above, so below. The land at Gateway has all the chakras of humans and animals. They are spread across the land top to bottom in order and recognizable by the physical clues they give us as in the layout of the land, how the trees grow, rock forms, etc. The Earth is speaking to us; the landforms, the trees, all tell their story. As I said earlier, we are dragon energy in this area, and the rocky backs of dragons stretch as far as the eye can see in our vast views here.

There are also many wood dragons, most of them gathered in our southwest quadrant where the Crown, Throat, Heart, and Third Eye chakras are situated. Last year I painted a large 'Winnie the Pooh' style map of the land and created a land walk book to lead people through the four quadrants, pinpointing the highlights of each area. I used to lead the afternoon walks, but now let people enjoy the area of their choosing at their leisure. I still lead a walk to the southwest quadrant so I can take pictures of people in the Third Eye chakra, sending their dreams to spirit. The third eye is a rock formation that is laid out on the land in the position of the skull on the human body. It is right where the forehead would be and also looks like you are sitting in the mouth of a stone dragon. A powerful chakra, its energy extends for about a 50-foot circle, and a dowsing or L-rod held in the middle, spins in a clockwise direction. There are many stories of people manifesting their dreams after sending them out to the world in the Third Eye. Another fascinating area is the southeast quadrant with the four directions and the cairn which was used by the natives to send smoke signals to any invading tribes

This area was the border between the Sushwap peoples to the north, which summered here, and the Okanagans to the south. Below the cairn is the beheading rock where I lost mine in my past native life here.

The beheading rock sits on a mound.

Some think a burial mound, but others believe there is a crystal city of light below. One of our first retreat helpers, Susan, used to see crystal children coming out of the earth from the holes that are below there, and in other places on the property. The Crystal Skull is at the head of the valley and from a distance has a face looking somewhat like a pharaoh. It faces south, and at an earlier workshop, we created a sacred circle around it, incorporating its surrounding north, east, and west rocks, and bits and pieces still remain.

The Root Chakra, just up the hill from the Guesthouse, has some powerful areas like the Portal, created when rocks are separated from each other, which takes you out of time as the energies from the rocks connecting to each other is very disorienting. The trees in the Root Chakra are deeply rooted through the large rocks they've managed to grow out of, splitting them with their roots. The overall feeling in the area is being deeply rooted or grounded in the earth. There are also several vortices which are marked by ground juniper growing in large circles from the energies. Everywhere there are rock spines, like vertebrae on the back of dragons telling stories of the underlying purpose of the land, metamorphosis, and manifestation.

My other favorite quadrant (actually they are all my favorite) is the northwest with the Astral Tree

The subtitle of this book is 'From Confusion to Wisdom,' and I wanted to share some of the wisdom I have gained over my almost 70 years in the 'Earth Program'. During the retreats, we offer a visualization given to us by the Astral Tree, and ancient twin tree in the northwest quadrant, deep in an area that we call Sherwood Forest. The visualisation invites you to journey high above Gateway to the stars, and a vast tunnel of light that leads to our home in spirit, our home between lives. It is there where our higher selves and our group souls and families meet again and again between journeys.

One of my between lives homes that I often go to is a Crystal Merkabah (like two pyramids, one inverted on top of the other and twisted slightly to form a star). I enter this structure to find 'Crystal Wizard' (my higher self), and as I walk out behind the structure, there is a beautiful meadow full of all the animals I have ever known and loved. They all come to greet me and welcome me back home. In another area, a beautiful green garden with radiant trees and flowers, my friends and relatives often gather and are all available in a thought. In fact, I can visit any of the places in spirit with just a focus (like L.E. teaches focus and intent).

The Akashic records or library, the Healing Temple (the enormous one we mimicked in our physical world pyramid-roofed version). The Tapestry, which is a woven structure of great beauty and myriad colored threads including metallic and iridescent versions stretching through time and space and comprised of the threads of every being (Delores Cannon talks about this in her books).

Whatever you focus on instantly manifests like the white room I met with the Ascended Masters and Christ in, and it is available for a revisit in a thought. Crystal Wizard is the being I most enjoy spending time with, seeing that he is myself in its highest expression. He has multicolored crystal eyes that flash those colors and manifest whatever they focus on. He has a wand and cloak to help him make things appear and disappear and shoulder-length silver, actually beyond a silver to iridescent silver-white hair (a more stunning version of mine at age 70).

That vision is like a Disney version for the human mass mind consciousness. Everything is possible there in an instant like in quantum physics, when we focus on the wave and it becomes the point (and it doesn't exist until we give it focus). As above so below, and we're just beginning to understand these laws of the universe at this time of our evolution of consciousness. I have come to know that **everything** is possible. We are co-creators with God, and the only limits are those that are subject to our own belief. How do I know what I know? The horses told me: or in reality, immersion in their energy and the energies of the natural world, the trees, plants, and all of natural creation, tunes us to a higher vibration; the same way a musical note is held by a tuning fork and tunes all the instruments around it.

Horse therapy has been around for decades now, and Heart Math among the other research is backing up the hearsay with science. Books are appearing on forest bathing, the secret lives of trees, and their importance to our very existence (if we run out of oxygen we're all done). Everything in the natural world is symbiotic, and beyond those individual relationships, there is the grander scale of life within life, the discovery of how important the microbiome is to our health (our little friends that live on and in planet animal).

Then there are the inhabitants of the macrobiome or planet earth (us all) all evolving together and growing from the seed to become the flower of God. Worlds within worlds all coming into being in focus and intent. We humans are just beginning to discover our limitations have held us to a 3D perspective (observer us observing polarity, the two dimensions which give us context in our world). We are moving beyond that limited perspective to many higher dimensions or points of observation, as many as the sacred geometry of the universe. We can view ourselves from any perspective, 5D on up, and it's not linear.

Time is the fourth dimension and is not linear, and again is dependent on focus. As the horses share in many of their books, humans create the rules for being human. We are very limited at this moment in our mass consciousness. When we begin to resonate at higher and larger levels of consciousness by connecting to other consciousness like those in the natural world and allow ourselves to be tuned by the higher vibration of animals and nature, our consciousness is expanded to eventually become the all-inclusive wisdom and knowing of One Consciousness. From there, what the horses call 'The All', all is available.

From Life Unbridled: quote (from the animals)

"Imagine, you humans, being able to see beyond and outside of your individual perspective. You would still enjoy occupying your own particular body, but you would not be limited by it. When not wholly immersed in time, space, and substance, you could float freely in the world of consciousness. Imagine you are in a meadow, a horse within a herd, overlooking a lake, on a beautiful sunny day. Not only do you experience yourself in that meadow, but you can also, leave your body in consciousness and become the meadow—each grass and flower bending with the wind, each creature living there, the fragrance and the feeling and the texture of the meadow itself. Then you can float upon the lake, feel the silkiness of the water, become the body of the lake wrapping yourself around the land.

Then you are the mountains, majestically climbing to their lofty summit, looking down upon the valley, surrounding the valley with your mighty arms. Floating up into the clouds, you continue on your journey, leaving the earth to visit other friends in other places.

Then, suddenly, something catches your attention: A deer comes into view, then bounds away, and immediately your spirit is with the deer, leaping across the meadow. You are overwhelmed with the urge to do the same, and instantly, simultaneously, the herd is off.

You are at once among them, running as one horse, as well as running like the herd itself, your consciousness spanning the timeless experience of horse energy running: You are the wind, you are freedom and exhilaration; you are the expression of the total experience—the meadow, mountains, sky, and lake.

You are the day itself and the joy of living.

That is how we animals experience life. We immerse ourselves in total experience here on this plane, within and without our physical selves, and in a thought, we can travel beyond the veil of this one life, one time, to all life and all existence. We experience the world through our individual expression but we are not trapped by it, and if it becomes uncomfortable to stay, we move easily to another existence, moving our consciousness as you move from one home to another. We do not get attached as you do, knowing that there is not just one life, but continual life. We will always resonate with what we love.

It is consciousness that is life, and that exists forever; bodies come and go.

Humans are heavily invested in their bodies because they have created their own separate laws, which limit their consciousness within the boundaries they have set.

However, it is no different from the time they believed that the world was flat, and they could fall off the edge. They have created imaginary edges, which can cause them no harm. The only way to change mass belief is to increase your mass to a larger, more all-encompassing mass belief.

We are here to help you celebrate life without limitation — life lived in spiritual freedom. Welcome us as siblings, entertain our differences and understanding. We are working toward a new world order, one where we all experience our heritage on this and all planes, as children of the ALL, heir to unlimited and infinite freedom of expression: LIFE (Love In Finite Expression), forever."

The God experiences I've had throughout my life range from an experience of God within as individual and myriad as all creation, we being creation itself, to being transported to experience beyond embodiment, beyond the Disney world of human afterlife, to floating weightless in a sea of love, being love itself. Here is my poem for those of you who missed it the first time in 'One With The Herd' and I will leave you readers to take a 'probiotic' and digest this latest sharing:

SEA OF LOVE

Floating, weightless
In a sea of love,
I have no boundaries,
Within and without
An indescribable joy,
Flows to the eternal reaches of my soul.
All that I am,
The simple truth,
I now remember.
And wanting never to forget,
Once again I flounder
Like a fish on shore.
Yet the memory lingers
And I want always
To return to the sea.

Chapter Eighteen

ZANY AND SCARY IDEAS (OR; A FEW LOOSE ENDS)

During my many quiet hours both in my youth and as a single mom, I also had time for invention (Possibly that Franklin gene again sharing its cellular memory).

When I was still living with my parents, I remember inventing the 'Love Box.' It was around the time that 'pet rocks' had their 10 minutes in the sun. I was newly returned from London, and the Beatles were in their prime with their Apple headquarters a couple of blocks from our Baker Street flat. We would see them drive by in their paisley covered Rolls Royce; "All you need is love." I could see it blasted through the store and people flocking to buy (instead of roses) for their sweetheart a 'Box of Love.' They were designed by me, produced by my indulgent father, inscribed with all kinds of witty product descriptions "100% pure, no additives or preservatives, keep carefully in a safe place" Of course, you can't see, touch or taste love, so the box was ostensibly empty. It was a big seller in theory, but they sat at my dad's warehouse as I did not know the first thing about marketing. That seemed to be the recurrent thread that ran through my ideas, probably even my tapestry on the spirit side.

Another idea that I came up with as a single mom was 'Gummers.' I had school-age children, and the school insisted on a pair of indoor shoes to be kept at school, which involved another pair of boots to go back and forth. It was bulky, something to lose in the translation and expensive. Gummers would be the solution. I tested some prototypes made by painting liquid latex in layers on the mold (a rubber deck shoe) as they only needed to be that height. They were very like their plastic hole-filled mimicry of today 'Crocs' without holes, of course. They were also multicolored with the colors and scents of bubble gum, (hence Gummers) and even glow-in-the-dark versions. They were no-slip, easy to apply, and remove, just like a half sock, lightweight, easy to carry, etc. etc. I tested the prototypes, and all the kids loved them, and the mothers wanted them.

Again the problem was marketing, finding a production company, or just selling the idea to a shoe manufacturer. All of those actions involved my time and money, none of which I had.

I still have a prototype in a drawer today, which I recently showed Joe (of gardening, odd job, and junk removal fame) who is also an inventor. Joe invented Bat-Guard, which is a brilliant plastic hand cover for children's hands when they're holding the bat during baseball to prevent hand injuries from the pitcher (apparently it happens frequently). It just slides on the bat and could even be used in the major leagues.

Again, finding someone to produce, or even more problematic market, is always the deterrent. Joe also thought of a running shoe cover to protect expensive running shoes for joggers, which is why I shared my Gummers idea. Just one more invention I wanted to mention, the pet toilet. For all those poor pets (dogs and cats, not horses) who are left alone for long hours while their owners are at work. It would be an installation in bathrooms like a shower base with a sensor to flush after use. Super-simple, why don't we have them in all homes? Mind you; it might give owners an excuse to sit in front of the screen rather than get outside and walk themselves along with their pets. I will never forget the visual on one of my favorite films of all time Wall-E, of the 'Pillsbury dough people' with their permanent nearsighted vision due to excessive time in front of a screen, being transported on floating conveyors because they could no longer walk.

Enough on inventions, just a few thoughts on screens and smart electronics and eventually A.I. All of these things scare me having realized that our connection and higher inspiration are dependent on our time spent in immersion with the natural world. I don't even use a cell phone (I can't say I don't have one as Kevin brought one home, but it spent all of its time on the floor of my truck with dead batteries until I stashed it in the bottom drawer somewhere inside the house. It definitely saves money on the phone plan)!

I can see the world moving towards the predictions of Aldous Huxley and George Orwell. Soon they will make microchips and vaccinations mandatory, but they'll never find me. Kind of like they used to demand, we fill out a census form, and I would pick it up with my mail and leave it on the floor of my truck until "oops, it slid out the door." Who are 'they' anyway, and how is my personal information any business of theirs? Actually, what happened to those forms, and how do 'they' get their information today? If you don't have a smart device, how do they spy on you? It's really creepy. I suppose if you can't replace your fridge or, in fact, an electrical appliance without inviting that creepy surveillance into your home, there will be no choice. Instead of us having harmless little house guests like mice and flies, you will be host to an army of A.I. And the humans hired to transcribe the information or giggle at us in our underwear.

There are so many human mass mind consciousness beliefs that keep us trapped in 3D, trapped in a belief in linear time

I have discovered that an intensity of focus shortens the distance between intention and fruition of an idea. I first started noticing it in my art. Years ago, I painted photorealism in my wildlife and nature art, painting every hair on a furry creature and every blade of grass in its surroundings. One would assume that painting this way with a fine liner brush and many layers of paint would be similar to needlepoint and take the time it took for completion. Not so. I found that focus or intention could be magnified both by physically executing the art while intensely focussing and also by simply holding it in focus between physical execution. I could simply transcend the process by knowing it was possible to speed up the time required. People would always ask me how long it took to paint a particular piece, and I would explain that I didn't keep track as each piece had a life of its own. I was given the subtle intention or inspiration and then gave over to higher consciousness its evolution, knowing that transcendence would complete the project in record time following my belief. It truly worked, and art that used to take me three weeks was completed in several days.

I apply the same principle to all subtle intention or inspiration that appears to me. I am grateful, excited, focused on any further evolution of the idea but allow it to transcend my limited understanding and gestate in the unlimited womb of creation.

From my cat Nune

I Am the darkness of possibility,
Warm and comforting.
I am the softness of a whisper,
Speaking without words.
I am truth gestating
in the womb of life,
Holding for you the memory of magic.
- Nune

As well as discovering time could be relative, I also discovered space and dimension were also flexible. I started by believing that I could manifest things with my focus and tested the theory on simple things like food items in my home. I would begin by needing a particular item, looking for it in all the possible places, and when not finding it, stop and rethink the situation. I knew I could manifest it!

Something like a jar of dill pickles or a can of soup would not be present after a thorough search. I would even send Kevin to back me up. Then after changing my belief to "I will find it," I would continue my search, and there it would be. It was a test that I used not only for things I needed or lost but also for Kevin's ever-elusive articles like his glasses, wallet, or keys. I would give him a day, maybe even two, and then ask where the object was, and there on a shelf that was suddenly eye-level to my search, it would suddenly materialize. I enjoy my experiments with matter and the laws currently accepted by the majority of humans. Really the possibilities are endless if you understand that thoughts and words become things. We are powerful manifestors when we know that to be true!

Another thing I've noticed is levels of consciousness or groups of consciousness that we subscribe to. Human Mass mind consciousness is heavily subscribed by humans. It is rife with shortcomings and tainted with timeworn dogma. The least reliable way in the world to tell what's happening is by listening to what people say.

People change their stories; people lie, people hide things, people agree to private clubs based on religion, location, and ethnicity.

They have agreed upon limitations like 'there is not enough' 'It's you against me.' Due to my isolation and saturation living in my heart space, in the moment, in the company of nature and my animals, I have been tuned to a version of human-mass-mind-consciousness diluted by One consciousness, which is a repository of all individual consciousness groups in creation. I am no longer a 'pure Raza Humana,' and I much prefer it that way.

Sometimes I feel completely disembodied, a point of consciousness just looking in on all the goings-on in this particular time, space, and dimension. Always I have a knowing that 'we' are so much larger, wiser, and evolved, like a futuristic version looking in on our earlier version. I guess that must be glimpses of God Consciousness. I absolutely trust in our evolution from this moment in our gestation to our birth as fully realized children of God. I also know we need to meld with all aspects of creation to be that All-knowing, All-encompassing, All-Loving, BEING.

The world is really at a crossroads, and I'm hoping that I'm too old to see that twin planet develop and float off to its own hell. I will be with the world of Light, of Wisdom, Truth, Joy, Freedom, Abundance, and Love. L.I.F.E (Love In Finite Expression) forever along with all the animals, trees, flowers, birds, and bees, the land of clean water and air (GMO and EMF free) and the only one who will have privilege to my innermost thoughts will be One MIND, God, or The All as my animals call it, and all those other enlightened beings that are One.

Nune

Imy, Kaia and Rhea, all grown up

Lunch Time at the Barn (main) House

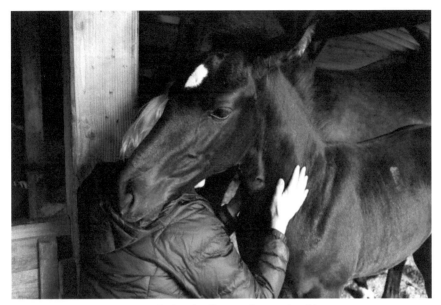

Moonshadow healing at a few weeks old

Connie with her boys

Liz and Amora

Peter (grown up?) and his dog Asher playing with Zac

The Proctor Family (Alex) with Granny and Grandpa

Sara and Amora

Jennifer and the cows

Pintos Unite

Sara and baby Ellie

October Color

December at the Ranch

Liz in the Solar Plexus Chakra in October

Liz riding in the woods in Gibsons